Technik im Fokus

Die Buchreihe Technik im Fokus bringt kompakte, gut verständliche Einführungen in ein aktuelles Technik-Thema.

Jedes Buch konzentriert sich auf die wesentlichen Grundlagen, die Anwendungen der Technologien anhand ausgewählter Beispiele und die absehbaren Trends.

Es bietet klare Übersichten, Daten und Fakten sowie gezielte Literaturhinweise für die weitergehende Lektüre.

More information about this series at http://www.springer.com/series/8887

Klaus Mainzer

Artificial intelligence – When do machines take over?

 Springer

Klaus Mainzer
Emeritus of Excellence
Technical University of Munich
Munich, Germany

Carl Friedrich von Weizsäcker
Center, Eberhard Karls University
of Tübingen, Tübingen, Germany

ISSN 2194-0770   ISSN 2194-0789   (electronic)
Technik im Fokus
ISBN 978-3-662-59716-3   ISBN 978-3-662-59717-0   (eBook)
https://doi.org/10.1007/978-3-662-59717-0

Translation from the German Language edition: Mainzer: Künstliche Intelligenz –
Wann übernehmen die Maschinen?, © Springer-Verlag GmbH Deutschland 2018

This Springer imprint is published by the registered company Springer-Verlag
GmbH, DE part of Springer Nature.
The registered company address is: Heidelberger Platz 3, 14197 Berlin, Germany

Foreword

Artificial intelligence has long dominated our lives without many people being aware of it. Smartphones that talk to us, wristwatches that record our health data, workflows that organize themselves automatically, cars, airplanes and drones that control themselves, traffic and energy systems with autonomous logistics or robots that explore distant planets are technical examples of a networked world of intelligent systems. They show us how our everyday life is determined by AI functions.

Biological organisms are also examples of intelligent systems which, like humans, have evolved during evolution and can solve problems more or less independently and efficiently. Occasionally nature is a model for technical developments (for example, neural networks as simplified models of the human brain). However, sometimes, computer science and engineering find solutions that are different, even better and more efficient than in nature. Therefore, there is not "the" artificial intelligence, but degrees of efficient and automated problem solving in different domains.

Behind this is the world of machine learning with learning algorithms, which become more and more powerful with exponentially growing computing capacity. Scientific research and medicine are already using neural networks and learning algorithms to discover correlations and patterns in a growing flood of measurement data. Machine learning algorithms are already applied in business strategies and the industrial Internet. They control the processes of a networked world in the Internet of

Things. Without them, the flood of data generated by billions of sensors and networked devices would not be manageable.

But, the state of the art in machine learning is based on statistical learning and reasoning with an exploding number of parameters. In general, statistical correlations cannot be replaced by causal explanations. They deliver risky black boxes trained by Big Data. Therefore, statistical learning must be overcome by causal learning. Causal learning does not only enable better explanation of causes and effects, but also better accountability to decide legal and ethical questions of responsibility (e.g. in autonomous driving or in medicine). Obviously, in addition to the innovation of artificial intelligence, the challenges of security and responsibility come to the fore. This book is a plea for certification and verification of AI-programs. We analyze empirical test procedures as well as automated formal proofs. In the end, the demand for certification is no killer of innovation, but the chance for better and sustainable AI-programs.

Since its inception, AI research has been associated with great visions of the future of mankind. Is "artificial intelligence" replacing humans? Some already speak of a coming "super intelligence" that triggers fears and hopes. This book is also a plea for technology design: AI must prove itself as a service in society. As service system, AI-technology with its immense need of energy must not be at the expense of ecology. Therefore, we should integrate the advantages of biological brains with their low rates of energy in new neuromorphic computer architectures. Quantum computing will also offer new computational technologies for AI.

Artificial intelligence is already a key technology that will determine the global competition of societal systems. The wealth of nations will depend decisively on their power of AI-innovation. But, their way of life will depend on their evaluation of AI-Technology. Will our political systems change under the influence of a dominating AI-technology? How are we supposed to assert our individual freedoms in the AI world? Europe will have to position itself not only as a technical AI location, but also with its moral value system.

Since my early studies as student, I was fascinated by the algorithms that make artificial intelligence possible. We need to know their foundations in order to assess their performance and limitations. Surprisingly, this is an essential insight of this book, no matter how fast supercomputers are, they do not change the logical-mathematical foundations proven by human intelligence. Only on the basis of this knowledge, societal impacts can be assessed. For this purpose, we had already founded the Institute for Interdisciplinary Computer Science at the University of Augsburg at the end of the 1990s. At the Technical University of Munich, I was also head of the Carl von Linde Academy and, as part of the Excellence Initiative 2012, founded the Munich Center for Technology in Society (MCTS). In 2019, I was inspired by a research project of the Volkswagen-Stiftung on the topic "Can software be responsible?" As a member of a High Level Group (HLG) of the German Ministery of Economy and the DIN Commission, we work on an roadmap of AI-certification. In the thematic network of the German Academy of Science and Technology (acatech), there is also "Technology in Focus - Data Facts Background", as this new book series is called by Springer. As a long-time author at Springer publisher, I thank the publisher for the proven support of the English translation of the German 2nd edition 2019.

Munich   Klaus Mainzer
in June 2019

Contents

Introduction: What Is AI?

After the ringing of my alarm clock has shocked me a little rough, the familiar and friendly female voice of Susanne wishes me a good morning and asks me how I slept. Somewhat sleepy I inquire after my appointments this morning. Susanne reminds me of an appointment in Frankfurt at our branch. Friendly, but certainly reminds me of the exercise training prescribed by a doctor. I look at my wristwatch, which shows my current blood pressure and blood values. Susanne's right. I'd have to do something. Susanne and the alarm clock are in my smartphone, which I put in my pocket after showering and dressing and hurry to the car. Turned to the cockpit of my car, I briefly explain my destination. Now I have time for a coffee and read the newspaper relaxed. My car's heading for the freeway on its own. On the way, the car evades a construction vehicle. It complies with the traffic regulations in an exemplary manner and nevertheless makes better progress than some human drivers who want to be faster stressed with excessive speed, flashing lights and too short distances. People are just chaotic systems, I still think. Then I ask Susanne to give me market profiles of our products, which she filters out with Big Data algorithms at lightning speed. Arriving at the Frankfurt branch I have the car parked independently. Semiconductor production in our plant is largely automatic. Special customer requests can also be entered online in the sales department. The production then adapts itself independently to these special wishes. Next week I want to go to Tokyo and meet our Japanese business partner. I still have to ask him not to put me in one of the new robot hotels. The last time I checked in, everything was automatic, just like checking in at the airport. Even in the reception a friendly robot lady sat. With human service, it'll be a little more expensive. But here I am European "old-fashioned" and at least in my private life I prefer human affection …

© Springer-Verlag GmbH Germany, part of Springer Nature 2020
K. Mainzer, *Artificial intelligence – When do machines take over?*, Technik im Fokus,
https://doi.org/10.1007/978-3-662-59717-0_1

That wasn't science fiction scenario. These were AI technologies which are technically feasible today and which, as part of the field of computer science and engineering, can be developed. Traditionally, AI (Artificial Intelligence) as a simulation of intelligent human thinking and acting. This definition suffers from the fact that "intelligent human thinking" and "acting" are not defined. Furthermore, man is made the yardstick of intelligence, although evolution has produced many organisms with varying degrees of "intelligence". In addition, we have long been surrounded in technology by "intelligent" systems which, although they are independent and efficient but often different from humans in controlling our civilization.

Einstein has answered the question "What is time?" independent of human imagination: "Time is what a clock measures." Therefore, we propose a working definition that is independent of human beings and only depends on measurable quantities of systems [1]. To this end, we look at systems that can solve problems more or less independently. Examples of such systems could be organisms, brains, robots, automobiles, smartphones or accessories that we wear on our bodies (wearables). Systems with varying degrees of intelligence are also available at factory facilities (industry 4.0), transport systems or energy systems (smart grids) which control themselves more or less independently and solve central supply problems. The degree of intelligence of such systems depends on the degree of self-reliance, the complexity of the problem to be solved and the efficiency of the problem-solving procedure.

So there is not "the" intelligence, but degrees of intelligence. Complexity and efficiency are measurable variables in computer science and engineering. An autonomous vehicle then has a degree of intelligence that depends on its ability to reach a specified destination independently and efficiently. There are already more or less autonomous vehicles. The degree of their independence is technically precisely defined. The ability of our smartphones to communicate with us is also changing. In any case, our working definition of intelligent systems covers the research that has been working successfully for many years in computer science and technology under the title "Artificial Intelligence" and is developing intelligent systems [2].

▶ **Working definition** A system is called intelligent when it can solve problems independently and efficiently. The degree of intelligence depends on the degree of autonomy of the system, the degree of complexity of the problem and the degree of efficiency of the problem-solving procedure.

It is true that intelligent technical systems, even if they have a high degree of independent and efficient problem solving, were ultimately initiated by people. But even human intelligence has not fallen from the sky and depends on specifications and limitations. The human organism is a product of evolution that is full of molecularly and neuronally encoded algorithms. They have developed over millions of years and are only more or less efficient. Randomness often played along. This has resulted in a hybrid system of abilities that by no means represents "the" intelligence at all. AI and technology have long since overtaken natural skills or solved them differently. Think of the speed of data processing or storage capacities. There was no such thing as "consciousness" as necessary for humans. Evolutionary organisms such as stick insects, wolves or humans solve their problems in different ways. In addition, intelligence in nature is by no means dependent on individual organisms. The swarm intelligence of an animal population is created by the interaction of many organisms, similar to the intelligent infrastructures that already surround us in technology and society.

Neuroinformatics attempts to understand the functioning of nervous systems and brains in mathematical and technical models. In this case, AI researchers work like natural scientists who want to test models of nature. This can be interesting for the technology, but does not have to be. AI researchers often work as engineers who find effective solutions to problems independently of the natural model. This also applies to cognitive skills such as seeing, hearing, feeling and thinking, such as modern software engineering shows. Even in the case of flying, the technology was only successful when it had understood the laws of aerodynamics and, for example, had developed other solutions with jet aircraft than in evolution.

In Chap. 2, we begin with a "Brief History of AI," linked to the great computer pioneers of the 20th century. The computer was first taught to reason logically. The computer languages developed for this purpose are still used in AI today. Logical-mathematical reasoning leads to automatic proofs that help save computer programs. On the other hand, their analysis is connected with deep-seated epistemological questions of AI (chap. 3). However, general methods are not sufficient to solve specific problems in different specialist areas. Knowledge-based expert systems simulated diagnoses by doctors and analyses by chemists for the first time. Today, expert systems are part of everyday life in research and work, without still being called "artificial intelligence" (chap. 4). One of the most spectacular breakthroughs of AI are speech processing systems, since language is traditionally considered the domain of man. The tools used show how different technology and evolution can solve problems (chap. 5).

Natural intelligence originated in evolution. It therefore makes sense to simulate evolution sing algorithms. Genetic and evolutionary algorithms are now also being used in technology. (chap. 6). Biological brains not only enable amazing cognitive performance such as seeing, speaking, hearing, feeling and thinking. They also work much more efficiently than energy-guzzling supercomputers. Neural Networks and learning algorithms are intended to decipher these abilities (chap. 7). The next step is humanoid robots in a human-like form that works together with people at work and in everyday life. In a stationary industrial robot, the work steps are defined in a computer program. Social and cognitive robots, on the other hand, must learn to perceive their environment, to decide independently and to act. This requires intelligent software with sensor technology to realize this kind of social intelligence (chap. 8).

Automobiles are already referred to as computers on four wheels. As autonomous vehicles, they generate intelligent behavior that is intended to more or less completely replace the human driver. Which application scenarios are associated with this in traffic systems? Like the swarm intelligence in nature, intelligence is not limited to individual organisms. In the Internet of Things, objects and devices can be equipped with intelligent software interfaces and sensors to solve problems collectively. A current example is the industrial Internet, in which production and sales are largely organized independently. A factory then becomes intelligent according to our working definition. In general, one speaks meanwhile of cyberphysical systems, smart cities, and smart grids (chap. 9).

Since its inception, AI research has been associated with great visions of the future of mankind. Will there be neuromorphic computers that can fully simulate the human brain? How do analogue processes of nature and digital technology differ? Will the technologies of artificial life converge with artificial intelligence? The book discusses new research findings on logical-mathematical fundamentals and technical applications of analog and digital techniques.

Despite the sobriety of everyday AI research, hopes and fears motivate and influence the development of high-tech societies. Especially in the strongholds of American information and computer technology such as Silicon Valley, one believes in a singularity when AI will replace humans. We are already talking about a collective superintelligence.

On the one hand, superintelligence, as shown in this book, would also be subject to the laws of logic, mathematics, and physics. We therefore need interdisciplinary basic research so that the algorithms do not get out of hand. On the other hand, we demand technical design: After the experiences of the past, we should recognize the chances, but also consider exactly for which purpose and use we should develop AI in the future. AI must prove itself as a service in society [2]. That is their ethical yardstick (chap. 10).

References

1. Mainzer K (2003) AI – Artificial Intelligence. Foundations of Intelligent Systems. Wissenschaftliche Buchgesellschaft, Darmstadt
2. DFKI (German Research Center for Artificial Intelligence). http://www.dfki.de/web. Accessed 8 Jan 2016

A Short History of the AI

2

2.1 An Old Dream of Mankind

An automaton is in ancient usage an apparatus that can act independently (autonomously). According to ancient understanding, self-activity characterizes living organisms. Reports on hydraulic and mechanical automats are already mentioned in ancient literature against the background of the technology of the time. In Jewish tradition, at the end of the Middle Ages, the Golem was described as a human-like machine. The Golem can be programmed with combinations of letters from the "Book of Creation" (Hebrew: Sefer Jezira)—to protect the Jewish people in times of persecution.

At the beginning of modern times, automation was approached from a technical and scientific point of view. From the Renaissance, Leonardo da Vinci's construction plans for vending machines are known. In the Baroque era, slot machines were built on the basis of watchmaking technology. P. Jaquet-Droz designs a complicated clockwork that was built into a human doll. His "androids" play the piano, draw pictures, and write sentences. The French physician and philosopher J. O. de Lamettrie sums up the concept of life and automata in the age of mechanics: "The human body is a machine that tensions its (drive) spring itself" [1].

The baroque universal scholar A. Kircher (1602–1680) already promotes the concept of a universal language in which

© Springer-Verlag GmbH Germany, part of Springer Nature 2020
K. Mainzer, *Artificial intelligence – When
do machines take over?*, Technik im Fokus,
https://doi.org/10.1007/978-3-662-59717-0_2

all knowledge is to be represented. Here, the philosopher and mathematician G. W. Leibniz directly follows and designs the momentous program of a "Universal Mathematics" (mathesis universalis). Leibniz (1646–1716) wants to trace thinking and knowledge back to arithmetic, in order to be able to solve all scientific problems by arithmetic calculations. In his age of mechanics, nature is imagined as a perfect clockwork in which every condition is determined as if by interlocking gears. Accordingly, a mechanical calculating machine executes each calculation step of a calculation sequence one after the other. Leibnizen's decimal machine for the four basic arithmetic operations is the hardware of his arithmetic calculations. Fundamental is the idea of a universal symbolic language (lingua universalis) in which our knowledge can be represented according to the model of arithmetic and algebra. What is meant is a procedure by which "truths of reason, as in arithmetic and algebra, can also be achieved, so to speak, by a calculus in any other area in which it is concluded" [2].

The further technical development from decimal calculators for the four basic arithmetic operations to program-controlled calculators did not take place in the scholars' room, but in the manufactories of the 18th century. There, the weaving of fabric samples is first controlled by rollers based on baroque slot machines, then by wooden punch cards. This idea of program control applies the British mathematician and engineer C. Babbage (1792–1871) on calculating machines. His Analytical Engine provided, in addition to a fully automatic calculation unit consisting of gears for the four basic arithmetic operations and a number memory, a punched card control unit, a data input device for numbers and calculation instructions, and a data output device with printing unit [3]. Although the technical functionality was limited, the scientific and economic significance of sequential program control in the age of industrialization is correctly recognized.

Babbage also philosophizes about analogies and differences between his calculating machines and living organisms and humans. His comrade-in-arms and partner Lady Ada Lovelace, daughter of the romantic poet Lord Byron, already prophesied:

"The Analytical Engine will process things other than numbers. When one transfers pitches and harmonies to rotating cylinders, this machine could compose extensive and scientifically produced pieces of music of any complexity and length. However, it can only do what we know to command it to do" [4]. In the history of AI, this argument of Lady Lovelace is mentioned again and again when it comes to the creativity of computers.

Electrodynamics and the electro-technical industry in the second half of the 19th century laid new technical foundations for the construction of computers. While Hollerith's tabulation and counting machine was being used, the Spanish engineer Torres y Quevedo thought about control problems for torpedoes and boats and constructed the first chess machine in 1911 for a final chess position with tower king vs. king.

Light and electricity also inspire writers, science fiction authors, and the beginning film industry. In 1923, the Czech writer Capek invented a family of robots. to free humanity from hard labor. After all, at least in the novel, the robots were provided with emotions. As machine men, they could no longer endure their slavery and rehearse the rebellion against their human masters. In the cinemas, movies like "Homunculus" (1916), "Alraune" (1918) and "Metropolis" (1926) were shown.

In industry and military research, the first special computers for limited computing tasks were built in the 1930s. However, the development of universal program-controlled computers, which can be programmed for different applications, will be fundamental for AI research. In April 11, 1936, the German engineer K. Zuse (1910–1995) applied for a patent for his "Methods for the automatic execution of calculations with the aid of calculating machines" [5]. In 1938, the Z1 was the first mechanical version to be completed, which was replaced in 1941 by the Z3 with electromechanical relay switches.

In 1936, for the first time, the British logician and mathematician A. M. Turing (1912–1954) defined the logical-mathematical concept of a computer: What is an automatic computational method, independent of its technical implementation? Turing's ideal computing machine requires an unlimited memory and only the smallest and simplest program commands, to which in principle any computer program, no matter how complicated, can be traced [6].

2.2 Turing Test

AI research in the narrower sense was born in 1950, when Turing published his famous essay "Computing Machinery and Intelligence" [7]. Here you will find the so-called "Turing Test". A machine is called "intelligent" if and only if an observer is unable to tell whether he is dealing with a human being or a computer. Observer and test system (human or computer) communicate via a terminal (today, e.g., with keyboard and screen). In his work, Turing presents sample questions and sample answers from various fields of application such as:

Example

Q   Please write me a poem about the Firth of Forth bridge.
A   I have to pass on this point. I could never write a poem.
Q   Add 34,957 to 70,764.
A   (waits about 30 s and then gives the answer) 105.721.
Q   Do you play chess?
A   Yes.
Q   My king stands on e8; otherwise I have no more figures. All they have left is their king on e6 and a tower on h1. It's your move. How do you draw?
A   (after a pause of 15 s) Th1-h8, matt.

Turing is convinced in 1950: "I believe that at the end of this century the general views of scholars will have changed to such an extent that one will be able to speak of thinking machines without contradiction". The fact that computers today calculate faster and more accurately and play chess better can hardly be denied. But people also err, deceive, are inaccurate and give approximate answers. This is not only a shortcoming, but sometimes even distinguishes them in order to find their way in unclear situations. In any case, these reactions should be able to be realized by a machine. The fact that Turing's test system did not want to write a poem, i.e. did not pass Lady Lovelace's creativity test, could hardly shake Turing. Which person is already creative and can write poems?

2.3 From "General Problem Solver" to Expert System

When in 1956 leading researchers like J. McCarthy, A. Newell, H. Simon et al. met at the Dartmouth conference on machine intelligence, they were inspired by Turing's question "Can machines think?" Characteristic was the interdisciplinary composition of this conference of computer scientists, mathematicians, psychologists, linguists and philosophers. Thus the group around the universally educated H. Simon, the later Nobel Prize winner for economics, advocated a psychological research program to investigate cognitive processes of human problem and decision making on the computer.

The first phase of AI research (around the mid-1950s to mid-1960s) is still dominated by euphoric expectations [8, 9]. Similar to Leibnizen's Mathesis Universalis, general problem-solving procedures are to be used for computers. After Newell, Shaw and Simon had developed the LOGICAL THEORIST in 1957, a proof program for the first 38 propositions from Russell's and Whitehead's logic book "Principia Mathematica", the GPS (General Problem Solver) program was to determine the heuristic basis for human problem solving at all in 1962. The disappointment was great given the practical results. The first specialized programs such as STUDENT for solving algebra tasks or ANALOGY for pattern recognition of analog objects proved more successful. It was found that successful AI programs depend on appropriate knowledge bases ("databases") and fast retrieval procedures.

In the second phase of the AI (around the mid-1960s to mid-1970s), an increased trend towards practical and specialized programming can be observed. Typical are the construction of specialized systems, methods for knowledge representation and an interest in natural languages. At MIT J. Moser developed the program MACSYMAL, which was actually a collection of special programs for solving mathematical problems in the usual mathematical symbolism. Further programs of this kind (e.g. for integration and differentiation) are still in practical use today.

In 1972, Winograd presented a robotics program to manipulate differently shaped and colored building blocks with a magnetic arm. For this purpose, the building blocks with their properties and locations were represented in data structures. Programming of the location information was carried out with the magnetic arm by changing the building blocks.

In the third phase of AI (around the mid-1970s to mid-1980s), knowledge-based expert systems that promised the first practical applications move to the fore. The delimited and manageable specialist knowledge of human experts such as engineers and doctors should be made available for daily use. Knowledge-based expert systems are AI programs that store knowledge about a specific field and automatically draw conclusions from that knowledge, in order to find concrete solutions or provide diagnoses of situations.

In contrast to the human expert, the knowledge of an expert system is limited. It has no general background knowledge, no memories, no feelings and no motivations, which can be different from person to person despite common special knowledge: An elderly family doctor who has known a family for generations will use different background knowledge in the diagnosis of a family member than the young specialist who has just left university.

Knowledge is a key factor in the representation of an expert system. We distinguish between two types of knowledge. One kind of knowledge concerns the facts of the field of application, which are recorded in textbooks and journals. Equally important is the practice in the respective area of application as knowledge of the second kind. It is heuristic knowledge on which judgement and any successful problem-solving practice in the field of application are based. It is experiential knowledge, the art of successful presumption, which a human expert only acquires in many years of professional practice.

E. Λ. Feigenbaum, one of the pioneers of this development, compared the development of knowledge-based expert systems in the mid-1980s with the history of the automotive industry. In the world of AI, it would be 1890, so to speak, when the first automobiles appeared. They were manually operated horseless cars, but already automobiles, i.e. self-driven vehicles. Just as Henry Ford had the first prototypes for mass production in his day, Feigenbaum also said that knowledge-based systems would go into mass production. Knowledge-based systems were thus understood as "automobiles of knowledge" [10].

References

1. de La Mettrie JO (2009) L'homme machine/The machine man. Meiner, Hamburg (translated and ed by C. Becker)
2. Leibniz GW (1996) Philos. Schr. VII. Olms, Hildesheim, p 32 (ed C.I. Gerhardt, repr.)
3. Babbage C (1975) On the mathematical powers of the calculating engine. In: Randell B (ed) The origins of digital computers—selected papers. Springer, Berlin, pp 17–52 (unpublished manuscript, Dec. 1837)
4. Lovelace C (1842) Translator's notes to an article on Babbage's analytical engine. Sci Mem 3:691–731 (ed V. R. Taylor)
5. Zuse K (1936) Procedure for the automatic execution of calculations with the aid of calculating machines. German patent application Z 23624 (April 11, 1936)
6. Turing AM (1936–1937) On computable numbers, with an application to the decision problem. Proc London Math Soc Ser 2(42):230–265
7. Turing AM (1987) Computing machinery and intelligence (1950). In: Turing AM (ed) Intelligence service. Fonts. Brinkmann and Bose, Berlin, pp 147–182
8. Feigenbaum EA, Feldman J (1963) Computers and thought. McGraw-Hill, New York
9. Mainzer K (1985) The concept of intelligence from the point of view of epistemology and science theory. In: Strombach W, Tauber MJ, Reusch B (eds) The concept of intelligence in the various sciences. Series of the Austrian computer society, vol 28. Oldenbourg, Vienna, pp 41–56
10. Further historical notes of AI in K. Mainzer (1994), Computer—new wings of the mind? The evolution of computer-aided technology, science, culture and philosophy. De Gruyter, Berlin, p 103 ff. The development of AI in the 1990s is documented in the AI journal "Artificial Intelligence" of the German Society of Informatics e. V

Logical Thinking Becomes Automatic

3

3.1 What Does Logical Reasoning Mean?

In the first phase of AI research, the search for general problem-solving methods was successful at least in formal logic. A mechanical procedure was specified to determine the logical truth of formulas. The procedure could also be executed by a computer program and introduced automatic proving in computer science.

The basic idea is easy to understand. In algebra, letters x, y, z… are used by arithmetic operations such as add (+) or subtract (−). The letters serve as spaces (variables) to insert numbers. In formal logic, propositions are represented by variables A, B, C…, which are connected by logical connectives such as "and" ($\land$), "or" ($\lor$)' "if-then" ($\rightarrow$), "not" ($\neg$). The propositional variables serve as blanks to use statements that are either true or false. For example, the logical formula $A \land B$, by using the true statements $1+3=4$ for A and $4=2+2$ for B, is transformed into the true statement $1 + 3 = 4 \land 4 = 2 + 2$. In arithmetic, this leads to the true conclusion $1 + 3 = 4 \land 4 = 2 + 2 \rightarrow 1 + 3 = 2 + 2$. But, in general, the conclusion $A \land B \rightarrow C$ is not true. On the other hand, the conclusion is $A \land B \rightarrow A$ is logically generally valid, since for the insertion of any true or false statements for A and B there is always a true overall statement.

The proof of the general validity of a logical conclusion can be very complicated in practice. Therefore, in 1965, J.A.

© Springer-Verlag GmbH Germany, part of Springer Nature 2020 15
K. Mainzer, *Artificial intelligence – When do machines take over?*, Technik im Fokus,
https://doi.org/10.1007/978-3-662-59717-0_3

Robinson proposed the so-called resolution method, according to which proofs can be found by logical refutation procedures [1–3]. One thus starts with the assumption of the opposite
(negation), i.e. the logical conclusion is not generally valid. In
the next step it is shown that all possible application examples
of this assumption lead to a contradiction. Therefore, the opposite of negation is true and the logical conclusion is generally
valid. Robinson's resolution method uses logical simplifications,
according to which any logical formula can be converted into a
so-called conjunctive normal form. In propositional logic, a conjunctive normal form consists of negated and non-negated propositional variables (literals), which are connected by conjunction
($\wedge$) and disjunction ($\vee$).

Example

For the conjunctive normal form $(\neg A \vee B) \wedge \neg B \wedge A$ the
formula consists of the clauses $A \vee B$, $\neg B$ and A, which
are connected by the conjunction $\wedge$. In this example, the literal $\neg A$ follows logically from the conjugation elements
$\neg A \vee B$ and $\neg B$. (The reason is simple: The conjunction
$B \wedge \neg B$ is always wrong for each application example for
B and $\neg A$ follows logically from $\neg A \wedge \neg B$). From $\neg A$ and the
remaining clause A, in the next step, the always wrong for-
mula $\neg A \wedge A$ follows, and thus the contradiction ε ("empty
word"):

Mechanically, therefore, the procedure consists of deleting con-
tradictory partial propositions from conjunctive elements of a
logical formula ("resolution") and repeating this procedure with
the resulting "resolvent" and another corresponding conjunctive
element of the formula until a contradiction (the "empty" word)
can be derived.

In a corresponding computer program, this procedure terminates for the propositional logic. Thus, it shows in finite time whether the presented logical formula is generally valid. However, the calculation time increases exponentially with the number of literals of a formula according to the previously known methods. Concerning "Artificial Intelligence", computer programs with the resolution method can automatically decide about the general validity of logical conclusions at least in the propositional logic in principle. People would have great difficulty keeping track of complicated and long conclusions. In addition, people are much slower. With increasing computing capacity, machines can therefore much more efficiently perform this task of logical concluding.

In predicate logic, statements are broken down into properties (predicates), which objects are assigned to or denied. Thus, in the statement $P(a)$ (e.g. "Anne is a student"), the predicate "student" (P) were assigned to an individual named "Anna" (a). This statement is either true or false again. In a predicative form of statement $P(x)$, blank spaces (individual variables) x, y, $z\ldots$ are used for the individuals a, b, $c\ldots$ of an assumed application domain (e.g., the students of a university). Beside logical connectives of propositional logic, now also quantifiers $\forall x$ ("For all x") and $\exists x$ ("There is a x") may be used. For example, $\forall x\, P(x) \rightarrow \exists x\, P(x)$ is a generally valid conclusion of predicate logic.

For the formulae of predicate logic, a generalized resolution procedure can also be indicated in order to derive a contradiction from the assumption of the general invalidity of a formula. For this purpose, a formula of predicate logic must be transformed into a normal form from whose clauses a contradiction can be derived mechanically. Since, however, in predicate logic (in contrast to propositional logic) the general validity of a formula cannot be decided in general, it can happen that the resolution procedure does not come to an end. The computer program then continues to run infinitely. It is then important to find subclasses in which the procedure not only terminates efficiently, but also at all. Machine intelligence can indeed increase the efficiency of decision-making processes and accelerate them. However, it is also (like human intelligence) restricted by the principle limits of logical decidability.

3.2   AI Programming Language PROLOG

To solve a problem with a computer, the problem must be translated into a programming language. One of the first programming languages was FORTRAN where a program consists of a sequence of commands to the computer like "jump to the position z in the program", "write the value a into the variable x". The focus is on variables, i.e. register or memory cells in which input values are stored and processed. Because of the commands entered, one also speaks of an imperative programming language.

In a predicative programming language, on the other hand, programming is understood as proving in a system of facts. This knowledge representation is well familiar from logic. A corresponding programming language is called "Programming in Logic" (PROLOG), which has been in use since the early 1970s [4–6]. The basis is the predicate logic, which we have already got to know in Sect. 3.1. Knowledge is represented in predicate logic as set of true statements. Knowledge processing is what AI research is all about. Therefore PROLOG became a central programming language of AI.

Here we want to introduce some modules of PROLOG in order to clarify the connection with knowledge processing. The logical statement "The objects $O_1, \ldots, O_n$ stand in the relation R" corresponds to a fact, which in predicate logic is given the general form $R(O_1, \ldots, O_n)$. In PROLOG you write:

$NAME(O_1, \ldots, O_n)$,

where "NAME" is any name of a relation. Strings that are represented in the syntactic form of facts are called literals.

Example

An example of a fact or literal is:

married (socrates, xantippe),
married (abélard, eloise),
is a teacher (socrates, plato),
is a teacher (abélard, eloise).

Statements and proofs about given facts now can be represented into question-answer-systems. Questions are marked with a question mark and outputs of the program with an asterisk:

? married (socrates, xantippe),

* yes,

? is a teacher (socrates, xantippe),

* no.

Questions can also refer specifically to objects for which variables are used in this case. In programming languages, descriptive names are used for this, such as "man" for any man or "teacher" for any teacher:

? married (man, xantippe),

* man = socrates,

? is a teacher (teacher, plato),

* teacher = socrates.

In general, a question in PROLOG is "are L_1 and L_2 and… and L_n" or in short:

? $L_1, L_2, …, L_n$

where $L_1, L_2, …, L_n$ are literals. Logical concluding rules such as the direct conclusion (modus ponens) are "if L_1 and L_2 and… and L_n is true, then L is also true" or in short:

$$L: - L_1, L_2, …, L_n$$

Example

This is how the rule can be introduced:

is a pupil (pupil, teacher):- is a teacher (teacher, pupil)

Then, it follows from the given facts:

? is pupil (pupil, socrates),

* student = plato

Based on a given knowledge base in the form of literals, PROLOG can find solutions to a question or problem using the resolution method.

3.3   AI Programming Language LISP

As an alternative to statements and relations, knowledge can
also be represented by functions and classifications such as those
used in mathematics. Functional programming languages therefore do not regard programs as systems of facts and conclusions
(such as PROLOG), but as functions of sets of input values in
sets of output values. While predicative programming languages
are involved in predicate logic, functional programming languages are based on the λ-calculus which A. Church defined in
1932/1933 for the formalization of functions with calculation
rules [7]. One example is the functional programming language
LISP, which was developed by J. McCarthy as early as the end
of the 1950s during the first AI-phase [8, 9]. Therefore it is one
of the oldest programming languages and was connected from
the beginning with the goal of artificial intelligence to bring
human knowledge processing to the machine. Knowledge is represented by data structures, knowledge processing by algorithms
as effective functions.

Linear lists of symbols are used as data structures in LISP
("List Processing Language"). The smallest (indivisible)
building blocks of LISP are called atoms. It can be numbers,
sequences of numbers or names. In arithmetic, the natural numbers are generated by counting, starting with the "atom" of one
(1) and then step by step the successor $n + 1$ by adding the one
from the predecessor number n. Therefore, arithmetic properties
are defined inductively for all natural numbers: You first define
a property for the one. In the inductive step, under the condition
that the property is defined for an arbitrary number n, it is also
defined for the successor $n + 1$ defined. Inductive definitions can
be generalized for finite symbol sequences. Thus, s-expressions
("s" for "symbolic") are formed from the atoms inductively as
objects of LISP:

▶ 1.   An atom is an s-expression.
   2.   If x and y are s-expressions, then also (x.y).

Examples of s-expressions are 211, (A.B), (TIMES.(2.2)), where 211, A, B, 2, and TIMES are considered atoms. Lists are now also defined inductively:

▶ 1. NIL ("empty symbol sequence") is a list.
   2. If x is an s-expression and y is a list, then the s-expression (x.y) is a list.

As a simplified notation, the empty list NIL is noted as () and the many brackets are written in the general form

(S1.(S2.(…(SN.NIL)…)))

simplified (S1 S2 … SN). Lists can again contain lists as elements, which allows the construction of very complex data structures.

Programming in LISP means algorithmic processing of s-expressions and lists. A function application is noted as a list, where the first list element is the name of the function and the remaining elements are the parameters of the function arguments. The following elementary basic functions are required for this: The CAR function, applied to an s-expression, returns the left part,

$(CAR(x.y)) = x,$

while the function CDR gives the right part,

$(CDR(x.y)) = y.$
The CONS function combines two s-expressions into one s-expression,

$(CONS\ x\ y) = (x.y).$

If you use these functions on lists, CAR returns the first element, CDR the rest of the list without the first element. CONS returns a list with the first parameter as the first element and the second parameter as the remainder.

Lists and s-expressions can also be displayed as binary ordered trees. Fig. 3.1a shows the tree representation of the general list (S1 S2 … SN) with the respective applications of the basic functions CAR and CDR, while Fig. 3.1b shows the s expression (A.((B.(C.NIL))). Function compositions such as (CAR.(CDR.x)) express the consecutive execution of two function applications, whereby the inner function is evaluated first. For multi-ary functions, all arguments are evaluated first and then the function is applied. Lists are usually regarded as the application of a function. Then (ABCDEF) means that function A is to be applied to B, C, D, E and F.

However, it often makes sense to interpret lists as (ordered) sets of symbols. So it makes little sense to read (14235) as application of function 1 to arguments 4, 2, 3, 5 when it comes to a sorting task of numbers. In LISP the symbol QUOTE is introduced, according to which the following list is not to be understood as a functional instruction, but as an enumeration of symbols: e.g. QUOTE(14235) or briefly `(14235). Then, by definition, B is CAR`(123) = 1, CDR`(123) = `(23), and CONS1`(23) = `(123). While variables are recorded by literal atoms, non-numeric constants can be distinguished by quoting variables: e.g., the variables x and LISTE and the constants `x and `LISTE.

According to these agreements, you use basic functions in LISP to define new functions. The general form of a function definition is as follows:

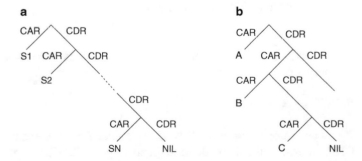

Fig. 3.1 Tree representations **a** the list (S1 S2 … SN) and **b** of the s expression (A.((B.(C.NIL))))

(DE NAME (P1 P2 ... PN) s-expression).

P1, P2, ..., PN are the formal parameters of the function, NAME is the designation of the function. The s-expression is the body of the function and describes the function application with the formal parameters. If in a program the function NAME appears in the form NAME (A1 A2 ... AN), then the formal parameters P1, P2, ..., PN must be replaced by the corresponding current parameters A1, A2, ..., AN in the body of the function and the body changed in this way must be evaluated.

Example

As an example, the function THREE is defined, which calculates the third element of a list:

(DE THREE (LISTE)(CAR(CDR(CDR(LISTE))))).

The function statement THREE`(415) replaces the formal parameters in the body of the function THREE with (CAR(CDR(CDR `(415))). The evaluation then returns the value 5 as the third element of the submitted list.

In order to be able to formulate conditions and case distinctions for functions, new atoms such as NIL for "false" and T ("true") and new basic functions such as EQUAL for comparing two objects:

(EQUAL 12)=NIL,
(EQUAL 11)=T.

The general form of a conditional expression in LISP is as follows:

```
(condition 1 s- expression 1)
(condition 2 s- expression 2)
          ⋮
(condition N s- expression N)
```

If the i-th condition ($1 \leq i \leq$ N) supplies the logical value T and all previous conditions supply the value NIL, the result of the conditional expression is the i-th s-expression. The conditional expression receives the value NIL if all conditions have the value NIL.

Example

As an example, a function is defined with which the length of a list can be calculated:

```
(DE LENGTH (LISTE)
   (COND
       ((EQUAL LISTE NIL)0)
         (T(PLUS(LENGTH
             (CDR LISTE))1)))))
```

The first condition determines whether the list is empty. In this case, it has the length 0. The second condition assumes that the list is not empty. In this case, the length of the list is calculated by adding the number 1 to the length of the list shortened by the first element (LENGTH(CDR LISTE)).

We now define what is generally known under a LISP-program.

▶ **Definition**
A LISP program is itself a list of function definitions and an expression to be evaluated with these function definitions:

```
((DE Funct 1 …)
(DE Funct 1 …)
    ⋮
(DE Funct N …)
  s- expression)
```

All previously defined functions can be used in the bodies of the used functions indicated by points. Since a LISP program itself is again an s-expression, programs and data in LISP have the same form. Therefore, LISP can also act as a meta-language for programs, i.e. LIPS can be used to talk about LISP programs. A further suitability of LISP for problems of knowledge processing in the AI is given by the fact that a flexible processing of symbols and structures is possible. Numerical calculations are only special cases.

AI attempts to structure problem-solving strategies algorithmically and then translate them into an AI programming language such as LISP. Search problems are a central AI application area. If, for example, an object is searched in a large quantity and no knowledge about a solution of the problem is available, then humans will also choose a heuristic solution, which is called British Museum algorithm.

Example

Examples for the British Museum algorithm are the search of a book in a library, a combination of numbers in a safe or a chemical formula under a finite number of possibilities under given conditions. The solution will certainly be found after this procedure, if all finally many possibilities or cases are examined and the following conditions are fulfilled:

1. There is a set of formal objects in which the solution is contained.
2. There is a generator, i.e. a complete enumeration procedure for this set.
3. There is a test, i.e. a predicate, which determines whether a created element belongs to the problem solving set or not.

The search algorithm is therefore also called "GENERATE_ AND_TEST (SET)" and should first be described in terms of content:

Function GENERATE_AND_TEST (SET)
If the SET quantity to be examined is empty,
then Failure,
in other respects
ELEM is the next element from SET;

If ELEM target element,
then deliver it as a solution,
otherwise repeat this function
with the SET quantity reduced by ELEM.

To formulate this function in LISP, the following help functions are required, the meaning of which is described: GENERATE creates an element of the given set. GOALP is a predicate function that returns T if the argument belongs to the solution set, otherwise NIL. SOLUTION processes the solution element for output. REMOVE returns the quantity reduced by the given element. The formalization in LISP is now as follows

```
(DE GENERATE_AND_TEST(SET)
        (COND ((EQUAL SET NIL)'FAIL)
          (T(LET(ELEM(GENERATE SET))
                 (COND((GOALP ELEM)(SOLUTION ELEM))
                 (T(GENERATE_AND_TEST
                        (REMOVE ELEM SET))))))))))
```

This example clearly shows that human thinking does not necessarily have to be a model for efficient mechanical problem solving. The goal is rather to optimize the human-machine interface with an expressive AI programming language. Whether these languages with their data structures also simulate or depict cognitive structures of human thought is a topic of cognitive psychology. AI programming languages are primarily used as computer-aided tools for optimal problem solving.

3.4   Automatic Proofs

If intelligence is to be realized with computers, then it must be traceable back to computing power. But, arithmetic is a mechanical process that can be broken down into elementary arithmetic steps. In arithmetic, elementary calculation steps can be performed by school children. We then speak of "algorithms" after the Persian mathematician al-Chwarismi who found solution methods for simple algebraic equations around 800 A.D. In 1936, Turing showed how a calculation process can be broken down into a series of smallest and simplest steps. Thus, for the first time, he succeeded in developing the general term of an effective method (algorithm) logically-mathematically. Only then is it possible to answer whether a problem is calculable in principle or not, regardless of the respective computer technology.

Turing imagined his machine as a typewriter. Like a movable writing head, a processor can print individual well-differentiated characters one after the other from a finite alphabet onto a strip tape (Fig. 3.2). The strip tape is divided into individual fields and is basically unlimited to the left and right. The program of a Turing machine consists of simple elementary instructions which are executed sequentially. Then, the machine can print or delete a symbol of the alphabet, move the read/write head one field to the left or to the right and finally stop it after many steps. Unlike a typewriter, a Turing machine can read the contents of individual fields of the tape one after the other and, depending on this, carry out further steps.

An example of a finite alphabet consists of the two characters 0 and 1, with which all natural numbers 1, 2, ... can be represented. As with counting, every natural number 1, 2, 3, 4, ... can be produced by adding 1 as often as necessary, i.e. 1, $1+1$, $1+1+1, 1+1+1+1$, ... A natural number is therefore

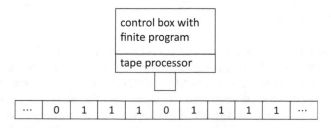

Fig. 3.2 Turing machine with strip tape

represented on the typewriter strip tape of a Turing machine by a corresponding chain of ones printed one after the other in a field. At the beginning and end, each number is limited with the digit 0. In Fig. 3.2, the numbers 3 and 4 are printed on a strip tape.

Example

An addition program for the task $3+4$ consists of deleting the zero between the two chains of ones and moving the left chain of ones one field to the right. Thus, a chain of seven ones is created, i.e. a representation of the number 7. The program then stops.

Not every program of a Turing machine is as simple as the addition. In principle, however, calculating with natural numbers can be traced back to manipulating 0 and 1 with the elementary commands of a Turing machine. In general, arithmetic investigates n-ary functions f with arguments $x_1, \ldots, x_n$ such as $f(x_1, x_2) = x_1 + x_2$. Each argument is a number represented on a Turing tape by a chain of digit 1. The remaining fields are empty, i.e. printed with 0. Now, the Turing computability of a function can be defined in general:

▶ At the beginning of a computation on a Turing tape there are only chains of ones, which are separated by zeros, thus: $\ldots 0x_1 0x_2 0 \ldots 0x_n 0 \ldots$ A n-ary function f with arguments $x_1, \ldots, x_n$ is called Turing-computable if and only if there is a Turing machine with a tape label $\ldots 0x_1 0x_2 \ldots 0x_n \ldots$ which, after finally many steps, stops with the tape marking $\ldots 0f(x_1, \ldots, x_n)0 \ldots$. The function value $f(x_1, \ldots, x_n)$ is represented by a corresponding chain of ones.

Each Turing machine can be uniquely defined by the list of its instructions. This Turing program consists of finitely many instructions and characters of a finite alphabet. Instructions and characters can be encoded by numbers. Therefore, a Turing machine can be uniquely characterized by a numerical code (machine number), which encrypts the corresponding machine program with its finite number of characters and arrangements. This machine number, like any number, can be recorded as a sequence of zeros and ones on a Turing band. Thus it was possible for Turing to simulate the behaviour of any Turing machine on any type of tape by a specific Turing machine with a given Turing tape. Turing called such a machine universal. It translates each instruction of the simulated machine, the machine code of which is noted on its tape, into a corresponding processing step of any given tape inscription.

From a logical point of view, any universal program-controlled computer, like e.g. built by John von Neumann or Zuse, is nothing else than technical realization of such universal Turing machine which can execute any possible Turing program. A computer today is a multi-purpose instrument that we can use as a typewriter, computer, book, library, video device, printer or multimedia show, depending on which program we set up and let run. But, smartphones and automobiles are also full of computer programs. In principle, each of these programs can be traced back to a Turing program. Because of their many tasks, these Turing programs would certainly be much more confusing, larger and slower than the programs we install today. But, from a logical point of view, these technical questions are irrelevant. In principle, every computer can compute the same class of functions that can also be computed by a Turing machine by increasing the memory capacity and extending the computing time at will.

In addition to Turing machines, various other methods for the definition of computable functions have been proposed, which prove to be mathematically equivalent to the computability by Turing machines [10, 11].

▶ In a thesis named after him (Church's thesis), A. Church stated that the intuitive concept of computability was completely covered by a definition such as Turing computability. Church's thesis cannot, of course, be proved, since it compares precise terms such as Turing computability with intuitive notions of computational procedures. However, it is supported by the fact that all previous proposals for definitions of computability are mathematically equivalent to Turing computability.

If we want to determine how intelligent a problem solution is, we must first clarify how difficult and complex a problem is. In the theory of computability, which goes back to Turing, the complexity of a problem is measured by the computational effort required to solve it. According to Church's thesis, the Turing machine is an exact measure of computability. The question of whether a problem can be decided effectively is directly related to its computability. For example, the question of whether a natural number is even or not can be decided in a finite number of steps by checking for a given number whether it is divisible by 2 or not. This can be computed with a program of a Turing machine.

However, it is not enough to be able to apply a given decision-making procedure to problems. It is often a question of finding solutions. We imagine a machine program that systematically lists all the numbers that solve a problem or fulfill a property.

▶ An arithmetic property is effectively enumerable if its fulfilling numbers can be enumerated (found) by an effectively computable method (algorithm).

In order to decide whether an arbitrarily presented number is even or not, it is not sufficient to effectively enumerate all even numbers one after the other in order to compare whether the number you are looking for is included. We must also be able to effectively enumerate all non-even (odd) numbers in order to compare whether the number we are looking for belongs to the set of numbers that do not meet the required property.

In general, a set is effectively decidable if it and its complementary set (whose elements are not part of the set) are effectively enumerable. Therefore, it follows that every effectively decidable quantity is also effectively enumerable. However, there are effectively enumerable sets that are not decidable. This brings us to the key question of whether there is also non-computable (non-algorithmic) thinking.

A example of a problem that cannot be decided effectively concerns the Turing machine itself.

▶ **Stop problem of the Turing machine** In principle, there is no general decision procedure for the question whether an arbitrary Turing machine with a corresponding machine program stops after a finite number of steps at an arbitrary input or not.

Turing began his proof of undecidability of the stop problem with the question whether all real numbers are computable. A real number like $\pi = 3,1415926 \ldots$ consists of an infinite number of digits after the decimal point, which seem to be randomly distributed. Nevertheless, there is a finite procedure or program for the step-by-step calculation of each digit with increasing accuracy by π. Therefore, π is a computable real number. In a first step, Turing defines a provable non-computable real number.

Background information

A Turing program consists of a finite list of symbols and operation instructions, which we can encrypt with number codes. In fact, this also happens in the machine program of a computer. In this way, each machine program can be uniquely characterized by a numerical code. We call this number the code or program number of a machine program. Now we imagine a list of all possible program numbers, which are listed in the order $p_1, p_2, p_3\ldots$ with increasing size. If a program computes a real number with an infinite number of digits after the decimal point (such as π), it is noted in the list after the corresponding program number. Otherwise the line behind a program number remains empty [12]:

$$p_1 \quad \cdot\cdot z_{\underline{11}}\, z_{12}\, z_{13}\, z_{14}\, z_{15}\, z_{16}\, z_{17} \cdots$$
$$p_2 \quad \cdot\cdot z_{21}\, z_{\underline{22}}\, z_{23}\, z_{24}\, z_{25}\, z_{26}\, z_{27} \cdots$$
$$p_3 \quad \cdot\cdot z_{31}\, z_{32}\, z_{\underline{33}}\, z_{34}\, z_{35}\, z_{36}\, z_{37} \cdots$$
$$p_4$$
$$p_5 \quad \cdot\cdot z_{51}\, z_{52}\, z_{53}\, z_{54}\, z_{\underline{55}}\, z_{56}\, z_{57} \cdots$$
$$\cdot$$
$$\cdot$$
$$\cdot$$

To define his non-computable number, Turing selects the underlined values on the diagonal of the list, modifies them (e.g. by adding 1) and assembles these modified values (with *) with a decimal point at the beginning to a new real number:

$$-.z^*_{11} \, z^*_{22} \, z^*_{33} \, z^*_{44} \, z^*_{55}\cdots$$

This new number cannot appear in our list, because it differs in the first digit of the first number after p_1 in the second digit from the second number after p_2, ... for all their digits after the decimal point. Therefore, the real number defined in this way cannot be computable.

With this figure, Turing proves in the next step the non-decidability of the stop problem. If the stop problem could be decided, then we would be able to decide whether or not a n-th decimal number behind the decimal point is computed, stopped and printed after a finite number of steps by the n-th computer program (with $n = 1, 2,...$). Therefore, we could compute a real number, which according to its definition cannot appear in the list of all computable real numbers.

Formal derivations (proofs) of the logic calculus can be understood as enumeration methods, with which the code numbers of the logical truths can be enumerated. In this sense, the set of logical truths of predicate logic of 1st order (PL1) is effectively enumerable. However, it cannot be decided effectively, since there is no general procedure for computing for any number whether it is the code number of a provable formula (i.e. a logical truth) of PL1 or not. It should be stressed that there is no general decision-making procedure for this calculus.

▶   **Important**
   The formal logic calculus of PL1 is complete, since we can use it to formally derive all logical truths of predicate logic of first order.
      In contrast, the formalism of arithmetic, with its basic arithmetic operations, is incomplete (K. Gödel's first incompleteness theorem, 1931).

In contrast to Gödel's extensive proof, the incompleteness of arithmetic follows directly from Turings stop problem:

Background information
If there were a complete formal axiomatic system from which all mathematical truths could be derived, then we would also have a decision procedure about whether a computer program would stop at some point.

We simply go through all kinds of proofs until we either find proof that the program is stopping, or find a proof that it is never stopping.

Therefore, if there were a finite set of axioms to derive all mathematical evidence, we could decide whether a computer program stops after a finite number of steps or not in contradiction to the non-decidability of the stop problem.

Gödel's second incompleteness theorem shows that the consistency of a formal system cannot be proved by the finite means used in the system itself. By finite evidence we mean such procedures that are modelled on the counting process 1, 2, 3, ...

If one extends the proof methods beyond the finite proof methods of this type, the consistency of the formal number theory becomes provable with stronger means. This was the basic idea of the logician and mathematician G. Gentzen (1909–1945), with which he introduced modern proof theory and gave important impulses for later computer programs in computer science. In 1936, the year of Turing's famous article on the decision-making problem, he wrote [13]:

"One can also express it in such a way that for the number theory no once and for all sufficient system of inferences can be indicated, but that rather propositions can be found again and again, whose proofs require novel kinds of inferences."

▶ Applied to computers, it follows that there cannot be "the" supercomputer that can decide all possible (mathematical) problems for arbitrary inputs. However, we can constantly add incomplete formalisms in order to arrive at richer and thus more powerful programs.

Background information
The complexity of formulas leads us to degrees of decidability [14]:

Thus the formula "For all natural numbers exists n and all natural numbers m there is a natural number p such that $m + n = p$" (formally: $\forall m \, \forall n \, \exists p \, m + n = p$) consists of an equation with the addition term m + n=p and variables *m, n* and *p* which is extended by an existential quantifier and two all-quantifiers. The addition of two numbers is effectively computable and therefore the property claimed in the equation is effectively decidable.

In general, an arithmetic formula consists of an effectively decidable property that is extended by logical quantifiers. Depending on the number, type and sequence of these quantifiers, classes of differently complex formulas can be distinguished that correspond to degrees of decidability.

Decidable properties correspond to Turing machines that stop after finally many steps. If quantifiers are added, the concept of the Turing machine must be extended, because computation processes may have to be run through several times infinitely (i.e. along all natural numbers). Sometimes, these processes are called "hypercomputation". However, these are only formal models of computing machines beyond the technical-physical realization by physical computing machines.

It is remarkable that Turing already addresses the topic of hyper-computability in his dissertation and asks how machines behave beyond effective algorithms.

▶   For Artificial Intelligence, the classes and degrees of
     computability and decidability are based on logical-
     mathematical proofs. They, therefore, are irrespective
     of the technical performance of physical computers.
     Even future supercomputers will not overcome the
     laws of logic and mathematics!

For the level of intelligence, the question of whether a problem can be decided in principle is not the only interesting aspect of solving a problem, but also how and with what effort a decision can be reached.

Example

Let's look at the well-known problem of a sales traveller who has to drive his customers in their different cities one after the other on the shortest possible route. With e.g. 3 customers it has 3 possibilities of the journey from home with the first customer. For the second customer, there are then $2 = 3 - 1$ possibilities of how to get there. For the third customer there is only $1 = 3 - 2$ possibility to get there and then drive back home. So the number of routes includes $3 \cdot 2 \cdot 1 = 6$ possibilities. Instead of possibilities the mathematicians say "faculty" and write $3! = 6$. With growing number of customers the possibilities increase extremely fast from $4! = 24$ over $5! = 120$ to $10! = 3.628.800$.

A practical application is the question of how a machine can drill 442 holes in a circuit board in the shortest way. Such printed circuit boards are found in similar numbers in household appliances, television receivers or computers. The number 442! with over a thousand decimal places cannot be tried out. How effective can a solution be?

In complexity theory of computer science, degrees of computability and decidability are distinguished [15, 16]. To do this, the computing time can be determined as the number of elementary steps of a Turing program depending on the length of the tape inscription at the beginning of computing ("input"). A problem has linear computing time if the computing time increases only proportionally to the length of the summands. In multiplication, the number of computation steps increases proportionally to the square of the input length. If the number of computation steps increases proportionally to the square, polynomial or exponent of the input length, we speak of quadratic, polynomial and exponential computational time.

Long computing times with a deterministic Turing machine are due to the fact that all partial problems and case distinctions of a problem must be systematically examined and computed one after the other. Sometimes it therefore seems more advisable to choose a solution from a finite number of possibilities by a random decision. This is how a non-deterministic Turing machine works: It randomly selects a possible solution to the problem and then proves the selected possibility. For example, in order to decide whether a natural number is composed, the non-deterministic machine selects a divisor, divides the given number with remainder by the divisor, and checks whether the remainder adds up. In this case, the machine confirms that the number is composed. On the other hand, a deterministic Turing machine must systematically search for a divisor by enumerating each number smaller than the given number and by performing the divisibility test.

▶ Problems decided in polynomial time by a deterministic machine are called P problems. If problems are decided by a non-deterministic machine in polynomial time, we speak of NP problems. According to this definition, all P problems are also NP problems. However, it is still an open question whether all NP problems are also P problems, i.e. whether non-deterministic machines can be replaced by deterministic machines with polynomial computing time.

Today we know of problems that are not decided, but of which we can determine exactly how difficult they are. From propositional logic, the problem is known how to find out which of the elementary propositions must be true or false such that the proposition composed of these elementary propositions is true (Sect. 3.1).

Thus the elementary propositions A and B connected by the logical connective "and" result in a true composite proposition, if and only if both elementary propositions A or B are true. The elementary propositions connected by "or" only result in a true composite proposition if and only if at least one elementary proposition is true. In these cases it is said that the composite proposition are "satisfiable". On the other hand, the elementary proposition A, which is connected by "and", and its negation do not form a satisfiable proposition: both elementary propositions cannot be true at the same time.

The computing time of an algorithm that checks all combinations of truth values of a composite proposition depends on the number of its elementary propositions. So far neither an algorithm is known which solves the problem in polynomial time, nor do we know whether such an algorithm exists at all. The American mathematician A. Cook, however, was able to prove in 1971 that the satisfiability problem is at least as difficult as any other NP problem.

▶ Two problems are equivalent in their difficulty if one solution to one problem also provides a solution to the other problem. A problem which in this sense is equivalent to a class of problems is called complete in relation to that class. After the satisfiability problem, other classic problems such as the problem of the traveling salesman can be proven to be NP complete.

NP-complete problems are considered hopelessly difficult. Practitioners are therefore not looking for exact solutions, but for almost optimal solutions under practicable restrictions. Inventiveness, imagination and creativity are required here. The round trip problem is just one example of the practical planning problems that arise every day when designing transport and communication networks requires algorithms under permanently changing network conditions. The less computation effort, time, and storage capacity are needed, the cheaper and more economical practical problem solutions are. The complexity theory thus provides the framework conditions for practical, intelligent problem solutions.

In AI, algorithms implement knowledge processing by deriving further series of characters from data structures. This corresponds to the ideal of a mathematical proof since antiquity. Euclid had shown how mathematical theorems could be derived and proved from axioms assumed to be true only by logical reasoning. In AI, the question arises whether mathematical proofs can be transferred to algorithms and "automated". Behind this is the basic AI question of whether and to what extent thinking can be automated, i.e. executed by a computer.

Let's take a closer look at a classic proof of this: Around 300 BC Euclid proved the existence of infinitely many prime numbers [17]. Euclid avoids the term "infinite" and claims: "There are more prime numbers than any number of prime numbers presented". A prime number is a natural number that has exactly two natural numbers as divisors. A prime number is therefore a natural number greater than one, which can only be divided by itself and by integer 1. Examples: 2, 3, 5, 7, …

Euclid argues with a proof of contradiction. He assumes the opposite of the assertion, concludes logically under this assumption on a contradiction. Therefore, the assumption was wrong. If we now assume that a statement was either true or false, then the opposite of the assumption applies: the statement is true.

Example

Now the proof of contradiction: Assuming there were only finally many prime numbers p_1, ..., p_n. With m we denote the smallest number which can be divided by all these prime numbers, the product $m = p_1 \cdot \ldots \cdot p_n$. For the successor $m + 1$ of m there are two possibilities:

First case: $m + 1$ is a prime number. By definition it is larger than p_1, ..., p_n and thus an additional prime number in contradiction to the assumption.

Second case: $m + 1$ is not a prime number. Then it must have a divider q. After assumption q must be one of the prime numbers p_1, ..., p_n. This also makes it a divider of m. The prime q therefore divides both m as well as the successor $m + 1$. Then it also divides the difference of m and $m + 1$, that is 1. However, this cannot apply, since 1 does not have a prime number as a divisor according to its definition.

The disadvantage of this proof is that we did not constructively prove the existence of prime numbers. We have only shown that the opposite of the assumption leads to contradictions. To prove the existence of an object, we need an algorithm that creates an object and proves that the statement for this example is correct. Formally a statement of existence reads $A \equiv \exists x\, B(x)$ (Sect. 3.1). In a diluted form, we could call for a list of finally many numbers $t_1, \ldots, t_n$ to construct the candidates for the statement B so that the OR-statement applies, i.e. the $B(t_1) \vee \ldots \vee B(t_n)$ statement is true for at least one of the constructed numbers $t_1, \ldots, t_n$. A machine could do that, too. If, in general, for all x a y with $B(x, y)$ should exist, i.e. formally $A \equiv \forall x \exists y\, B(x, y)$ is valid, then we need an algorithm p which constructs for each x-value a value $y = p(x)$ so that $B(x, p(x))$ for all x is valid, i.e. formally $\forall x B(x, p(x))$. In a weaker form we would be satisfied if for the search process of an y-value for a given x-value at least one upper barrier $b(x)$ could be calculated, i.e. formally $\forall x \exists y \leq b(x) B(x,y)$. This allows the search process to be precisely estimated.

The American logician G. Kreisel has therefore demanded that proofs should be more than mere verifications. In a way, they are "frozen" algorithms. You only have to discover them in the proofs and "wind them out" (unwinding proofs) [18]. Then they can also take over machines.

Example

In fact, a constructive procedure is "hidden" in Euclid's proof. For any position r of a prime p_r (in the enumeration $p_1 = 2$, $p_2 = 3$, $p_3 = 5,...$), an upper barrier $b(r)$ can be computed, i.e. one can specify a further prime number for each submitted number of prime numbers, which, however, lies below a computable barrier. In Euclid's proof many prime numbers are finally assumed, which are smaller or equal to a barrier x, so $p \leq x$. By that, we can construct a number $1 + \prod_{p \leq x} p$ from which the contradictions are derived. ($\prod_{p \leq x} p$ is the product of all prime numbers that are less than or equal to x are.) Therefore, we first construct the barrier

$$g(x) := 1 + x! \geq 1 + \prod_{p \leq x} p.$$

The faculty function $1 \cdot 2 \cdot \ldots \cdot x = x!$ can be estimated by the so-called Stirling formula. However, we are aiming at an upper barrier of the $r+1$-th prime number p_{r+1}, that only depends on the position r in the enumeration of prime numbers instead of the unknown barrier $x \geq p_r$. Euclid's proof shows $p_{r+1} \leq p_1 \cdot \ldots \cdot p_r + 1$. From that, we can prove for all $r \geq 1$ (by complete induction via r) that $p_r < 2^{2^r}$. Therefore, the computable barrier we are looking for is $b(r) = 2^{2^r}$.

In logic and mathematics, formulas (i.e. series of symbols) are derived step by step until the proof of an assertion is completed.

Computer programs basically work like proofs. Step by step, they derive strings according to defined rules until a formal expression is found that stands for a solution of the problem. For example, imagine the construction of a car on an assembly line. The corresponding computer program describes how the car is created step by step from presupposed individual parts according to rules, building on each other.

A customer wants a computer program from a computer scientist that solves such a problem. In a very complex and confusing production process, he certainly wants a proof beforehand that the program is working correctly. Possible errors would be dangerous or cause considerable additional costs. The computer scientist refers to a software that automatically extracts the proof from the formal properties of the problem. Just like software in "Data Mining" is used to search for data or data correlations, suitable software can also be used to automatic search for proofs. This is referred to as "proof mining" [19]. This corresponds to Georg Kreisel's approach to filter algorithms out of proofs (unwinding proofs), but now automatically by computer programs.

However, this raises the question of whether the software used to extract the evidence itself is reliable. Such proofs of reliability for the underlying software can be provided within a precisely defined logical framework. The customer can then be sure that the computer program is working correctly to solve the problem. This "automatic proof" is therefore not only of considerable importance for modern software technology. It also leads to philosophically deep questions, namely to what extent (mathematical) thinking can be automated: The search for proofs is automatic. However, the proof of correctness of the software used for this purpose is provided by a mathematician. Even if we were to automate this proof again, a fundamental epistemological question arises: Does this not lead us into a regress, at the end of which man always stands (must stand)?

One example is the interactive proof system MINLOG which automatically extracts computer programs from formal evidence [20, 21]. It uses the computer language LISP (Sect. 3.3). A simple example is the assertion that for each list v of symbols in

LISP a reverse list w exists with the symbols in reverse order. That's another assertion of the form $A \equiv \forall v \exists w \, B(v, w)$. The proof can be provided informally by an induction on the structure of the lists v. MINLOG automatically extracts a suitable computer program from it. But this software can also be used for demanding mathematical proofs. A general proof of reliability guarantees that the software delivers correct programs [22].

In 1969, the logician W.A. Howard observed that Gentzen's proof system of natural deduction can be directly interpreted in its intuitionistic version as a typed variant of the mode of computation known as λ (lambda) – calculus [23]. This basic insight of mathematical proof theory opened avenues to new generations of interactive and automated proof assistants.

According to A. Church, $\lambda a.b$ means a function mapping an element a onto the function value b with application $\lambda a.b[a] = b$. In the following, proofs are represented by terms $a, b, c, \ldots$; propositions are represented by $A, B, C, \ldots$. A proof of implication $A \to B$ (which means: if A, then B) starts with the assumption [A] of proposition A (assumption marked by brackets) and derived proposition B. This proof is understood as a function $\lambda a.b$ which maps an assumed proof a of proposition A onto a proof b of proposition B which is written in Gentzen-style:

$$
\lambda a.b \quad
\begin{array}{c}
[A] \\
\vdots \\
\dfrac{B}{A \to B}
\end{array}
$$

According to Church's thesis, the lambda-term $\lambda a.b$ represents a computer program. Following the Curry-Howard correspondence, a proof is considered a program, and the formula it proves is the type for the program. In the example, $A \to B$ is the type of the program $\lambda a.b$.

The calculus of constructions CoC is a type theory of Thierry Coquand et al. which can serve as typed programming language as well as constructive foundation of mathematics [24]. It extends the Curry-Howard correspondance to proofs in the full intuitionistic predicate calculus. CoC only needs very few rules of construction for terms. The objects of CoC are, e.g., proofs (terms with propositions as types), propositions, predicates (functions that return propositions), large types (types of predicates).

The calculus of inductive constructions CiC is based on CoC enriched with inductive definitions for constructing terms. An inductive type is freely generated by a certain number of constructors. An example is the type of natural numbers which is inductively defined by the constructors 0 and succ (successor). The type of finite lists of elements of type A is defined inductively with constructors nil and cons (cf. Sect. 3.3).

The proof assistant Coq implements a program which is based on the calculus of inductive constructions (CiC) combining both a higher-order logic and a richly-typed functional language [25]. The commands of Coq allow

- to define functions or predicates (that can be evaluated efficiently)
- to state mathematical theorems and software specifications
- to interactively develop formal proofs of these theorems
- to machine-check these proofs by a relatively small certification
- to extract certified programs to computer languages (e.g., Objective Caml, Haskell, Scheme)

Coq provides interactive proof methods, decision and semi-decision algorithms. Connections with external theorem provers are available. Coq is a platform for the verification of mathematical proofs as well as the verification of computer programs in CiC.

▶ A hardware or software program is correct ("certified by Coq") if it can be verified to follow a given specification in CiC.

Example

Coq is already applied in the fully automated Métro line 14 in Paris.

The structure and behaviour of electrical circuits can mathematically be modeled by interconnected finite automata.

In circuits, one has to cope with infinitely long temporal sequences of data (streams). A circuit is correct iff, under certain conditions, the output stream of the structural automaton is equivalent to that of the behavioral automaton. Therefore, automata theory must be implemented into CiC in ordert o prove the correctness of circuits [26].

What are the advantages of the Coq Proof Assistent for verification of software and hardware?

- In Coq, a verification of a computer program is as strong and save as a mathematical proof in a constructive formalism [27, 28].
- The use of Coq types provide precise and reliable specifications.
- The hierarchical and modular approach allows correctness results in a complex verification process related to pre-proven components.

There are further proof assistants like, e.g., Agda and Isabelle. From a practical point of view, the disadvantage is the increasing complexity of AI-programs which often cannot be handled by proof assistants with the accuracy of mathematical proofs. Later in this book, we come back to the challenge of security in modern machine learning.

References

1. Robinson JA (1965) A machine oriented logic based on the resolution principle. J Assoc Comput Mach 12:23–41
2. Richter MM (1978) Logikkalküle. Teubner, Stuttgart, p 185
3. Schöning U (1987) Logic for computer scientists. B.I. Wissenschaftsverlag, Mannheim, p 85
4. Kowalski B (1979) Logic for problem solving. North-Holland, New York
5. Hanus M (1986) Problem solving in PROLOG. Vieweg + Teubner, Stuttgart, Germany
6. Schefe P (1987) Computer Science—a constructive introduction. LISP, PROLOG and other programming concepts. B.I. Wissenschaftsverlag, Mannheim, p 285

7. Church A (1941) The Calculi of lambda-conversion. Library of America, Princeton (repr. New York 1965)
8. McCarthy J et al (1960) LISP 1 programmer's manual. MIT Computer Center and Research Lab. Electronics, Cambridge (Mass.)
9. Stoyan H, Goerz G (1984) LISP—an introduction to programming. Springer, Berlin
10. Hermes H (1978) Enumerability, decidability, computability. Introduction to the theory of recursive functions, 3rd edn. Springer, Berlin (1st ed. 1961)
11. Brauer W, Indermark K (1968) Algorithms, recursive functions and formal languages. B.I. Wissenschaftsverlag, Mannheim
12. Chaitin G (1998) The limits of mathematics. Springer, Singapore
13. Gentzen G (1938) The present situation in mathematical foundational research. Ger Math 3:260
14. Shoenfield JR (1967) Mathematical logic. Addison Wesley, Reading (Mass.)
15. Arora S, Barak B (2009) Computational complexity. A modern approach. Cambridge University Press, Cambridge
16. Wegener I (2003) Complexity theory. Limits of the efficiency of algorithms. Springer, Berlin
17. Aigner M, Ziegler GM (2001) Proofs from The book, 2nd edn. Springer, Berlin, p 3
18. Feferman S (1996) Kreisel's "unwinding" program. In: Odifreddi P (ed) Kreisleriana. About and Around Georg Kreisel, Review of Modern Logic, pp 247–273
19. Kohlenbach U (2008) Applied proof theory: Proof interpretations and their use in mathematics. Springer, Berlin (Chapter 2)
20. Schwichtenberg H (2006) Minlog. In: Wiedijk F (ed) The seventeen provers of the world. Lecture notes in artificial intelligence, vol 3600. Springer, Berlin, pp 151–157
21. Schwichtenberg H, Wainer SS (2012) Proofs and computations. Cambridge University Press, Cambridge (Chapter 7)
22. Mayr E, Prömel H, Steger A (eds) (1998) Lectures on proof verification and approximation algorithms. Lecture notes in computer science, vol 1967. Springer, Berlin
23. Howard WA (1969) The formulae-as-types notion of construction. In: Seldin JP, Hindley JR (eds) To H.B. Curry: essays on combinatory logic, lambda calculus and formalism. Academic Press, Boston, pp 479–490
24. Coquand T, Huet G (1988) The calculus of constructions. Inf Comput 76(2–3):95–120
25. Bertot Y, Castéran P (2004) Interactive theorem proving and program development. Coq'Art: the calculus of inductive constructions. Springer, New York
26. Coupet-Grimal S, Jakubiec L (1996) Coq and hardware verification: a Case Study. TPHOLs ,96, LCNS 1125:125–139

27. Mainzer K, Schuster P, Schwichtenberg H (2018) Proof and computation. Digitization in mathematics, computer science, and philosophy. World Scientific Publisher, Singapore
28. Mainzer K (2018) The digital and the real world. Computational foundations of mathematics, science, technology, and philosophy. World Scientific Publisher, Singapore

Systems Become Experts

4

4.1   Architecture of a Knowledge-Based Expert System

Knowledge-based expert systems are computer programs that store and accumulate knowledge about a specific area, from which knowledge automatically draws conclusions in order to offer solutions to concrete problems in that area. In contrast to the human expert, the knowledge of an expert system is limited to a specialized information base without general and structural knowledge about the world [1–3].

In order to build an expert system, the knowledge of the expert must be laid down in rules, translated into a program language and processed with a problem-solving strategy. The architecture of an expert system therefore consists of the following components: Knowledge base, problem-solving component (derivation system), explanatory component, knowledge acquisition, dialogue component. The coordination of this knowledge is the key factor in the representation of an expert system. There are two types of knowledge. One type of knowledge concerns the facts of the field of application, which are recorded in textbooks and journals. Just as important is the practice in the respective field of application as knowledge of the second kind.

© Springer-Verlag GmbH Germany, part of Springer Nature 2020   47
K. Mainzer, *Artificial intelligence – When do machines take over?*, Technik im Fokus,
https://doi.org/10.1007/978-3-662-59717-0_4

It is heuristic knowledge on which judgement and any success-ful problem-solving practice in the field of application are based. It is experiential knowledge, the art of successful presumption, which a human expert acquires only in many years of profes-sional work (Fig. 4.1).

The heuristic knowledge is most difficult to represent because the expert is usually not aware of himself. Therefore, interdis-ciplinary trained knowledge engineers have to learn the expert rules of human experts, present them in programming languages and convert them into a functional work program. This compo-nent of an expert system is called knowledge acquisition.

The explanatory component of an expert system has the task of explaining the examination steps of the system to the user. The question "How" aims at the explanation of facts or state-ments which are derived by the system. The question "Why" demands reasons for questions or commands of a system. The dialog component concerns the communication between the expert system and the user.

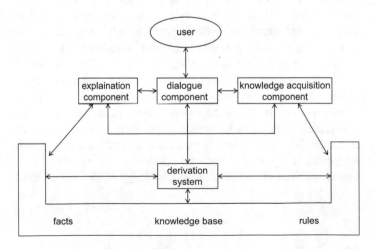

Fig. 4.1 Architecture of knowledge-based expert systems

4.2   Programming of Knowledge Presentations

A widespread representation of knowledge is rule-based. For application in expert systems, rules are understood as if-then statements in which the precondition describes a situation in which an action is to be executed. This can mean a deduction, according to which a statement is derived from a premise. An example exists when an engineer concludes from certain symptoms of an engine that an engine piston is defective. However, a rule can also be understood as an instruction to change a condition. If, for example, a piston is defective, the engine must be switched off immediately and the defective part replaced.

A rule-based system consists of a database with valid facts or states, rules for deriving new facts or states, and the rule interpreter for controlling the derivation process. There are two alternatives for linking the rules, which are called forward reasoning and backward reasoning in AI (Fig. 4.2; [4, 5]).

In forward chaining, one of the rules whose precondition is fulfilled by a given data basis is selected from the data basis, its action part is executed and the data basis is changed. This process is repeated until no more rules can be applied. The procedure is therefore data-driven. In a preselection, the rule interpreter first determines the systems of all executable rules that can be derived from the data basis as part of the respective

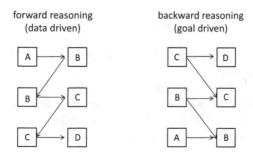

Fig. 4.2 Forward and backward reasoning

expert system. A rule is then selected from this set according to different criteria. A certain sequence, the structure of a rule, or additional knowledge can be decisive.

With backward chaining, only the rules whose action part contains the target are checked starting from a target. The procedure is therefore goal driven. If parts of the precondition are unknown, they are queried or derived with other rules. Backward chaining is particularly suitable if facts of the knowledge base are still unknown and must therefore be queried. The rule interpreter now begins with the given goal. If the target is unknown in the database, the rule interpreter must first decide whether the target can be derived or is to be queried. If a derivation is possible, all rules are executed in whose action part the target is contained. Unknown parts must be queried and derived as sub-targets.

A qualified expert has a complex basic knowledge which a structured data structure in an expert system must correspond to. For such a structuring of knowledge, all statements about an object are often summarized in a schematic data structure, which according to M. Minsky is also called "frame". A simple example for the summary of all properties of an object in a "frame" is [6]:

Object	Property	Value
Zebra	is a	mammal
	colour	striped
	has	hoofs
	size	large
	living space	ground

The properties are also called "slots", into which "filler", i.e. the values are entered.

Historically, Minsky uses linguistic templates for knowledge representation. Thus, an event such as the one in the sentence "Galilei observes Jupiter with the telescope" can be described by the following net-like framework:

From the central node "event" edges like "action", "actuator", "instrument", "target", "is on" go out and form the scheme of a semantic network, into which special objects like "Galilei", "telescope", "Jupiter" etc. are introduced. The directed edges correspond to the "slots" (properties) and the nodes to the "fillers" (values). The graphical notation of schemata by semantic networks obviously allows a clear representation of complex data structures.

In everyday life, cognitive schemata are activated in different situations. This can involve the recognition of typical objects, action in typical events or answers to typical questions. The respective "filler" of a concrete object are filled into the "slots" of the schema ("frame"). In the case of diagnostic tasks of a physician, for example, it may be necessary to classify concrete symptoms of a patient in a general "clinical picture" that is represented by a schema.

Relationships between objects are often represented by so-called "constraints". They are suitable for the representation of boundary conditions, with which the performance possibilities of a problem are limited. These can be constraints, e.g. when an engineer solves a technical problem, or constraints when preparing an administrative planning task. If the problem is mathematized, constraints are represented by mathematical equations or constraint networks by systems of equations [7].

Historically, DENDRAL was one of the first successful expert systems developed by E. A. Feigenbaum and others at Stanford in the late 1960s [8, 9]. It uses the special knowledge of a chemist to find a suitable molecular structural formula for a chemical sum formula. In a first step, all mathematically possible spatial arrangements of the atoms are systematically determined for a given sum formula.

For $C_{20}H_{43}N$ for example, there are 43 million orders. Chemical knowledge about the binding topology, according to which, for example, carbon atoms can be bound in multiple ways, reduces the possibilities to 15 million. Knowledge about mass spectrometry, about the most probable stability of bonds (heuristic knowledge) and nuclear magnetic resonance finally limits the possibilities to the desired structural formula. Fig. 4.3 shows the first derivation steps for the C_5H_{12}, for example.

The problem-solving strategy used here seems to be nothing more than the familiar "British Museum Algorithm", which is described in Sect. 3.3 in the LISP programming language. The procedure is therefore GENERATE_AND_TEST, whereby the possible structures are systematically generated in the GENERATE part, while the chemical topology, mass spectrometry, chemical heuristics and nuclear magnetic resonance each specify test predicates in order to restrict the possible structural formulae.

It is useful to divide problem solution types into diagnostic, design and simulation tasks. Typical diagnostic problem areas are medical diagnostics, technical diagnostics such as, e.g., quality control, repair diagnostics or process monitoring and object

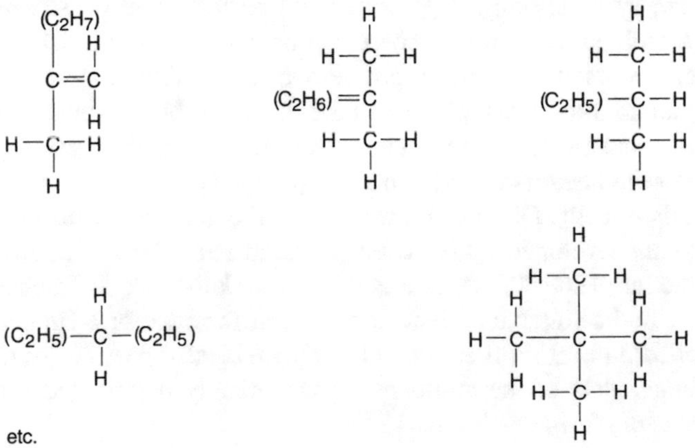

Fig. 4.3 Derivation of a chemical structural formula in DENDRAL

recognition. This is why DENDRAL also solves a typical diagnostic problem by recognizing the appropriate molecular structure for a given sum formula.

The first medical example of an expert system was MYCIN which was developed at Standford University in the mid-1970s [10, 11]. The MYCIN program was written for medical diagnosis to simulate a physician with medical expertise in bacterial infection. Methodically, it is a deduction system with back chaining. MYCIN's knowledge pool about bacterial infections consists of about 300 production rules. The following rule is typical:

Example

If the infection type is primary bacteremia, the suspected entry point is the gastrointestinal tract, and the site of the culture is one of the sterile sites, then there is evidence that the organism is bacteroides.

In order to be able to apply the knowledge, MYCIN works backwards. For each of 100 possible hypotheses of diagnosis, MYCIN tries to find simple facts that are confirmed by laboratory results or clinical observations. Since MYCIN works in an area where deductions are hardly safe, a theory of plausible reasoning and probability assessment has been linked to the deduction apparatus. These are so-called safety factors for each ending in an AND/OR tree, as the example in Figure Fig. 4.4.

In the tree, F_i denotes the safety factor that a user assigns to a fact. C_i indicates the safety factor of a termination, A_i the degree of reliability attributed to a production rule. Safety factors of the corresponding formula are calculated at the AND or OR nodes. If the safety factor of a data specification should not be greater than 0.2, it is considered unknown and is given the value 0. The program therefore calculates confirmation degrees depending on more or less safe facts. MYCIN has been generalized for various diagnostic applications independently of its specific database on infectious diseases.

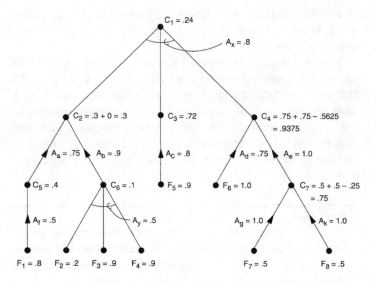

Fig. 4.4 AND/OR tree of a medical expert system

4.3   Limited, Uncertain, and Intuitive Knowledge

Experts are not distinguished by the fact that they can distinguish with absolute certainty between true and false and thus pretend to be more accurate than can be achieved. Rather, a good expert is able to assess uncertainties that arise, for example, in the medical diagnosis in the symptom assessment or symptom evaluation. In expert systems, therefore, classical logic is often not used with the assumption of the bivalence of the truth values ("tertium non datur"), but in addition uncertainty values such as "safe", "probable", and "possible" are assumed. It is an old problem in the theory of science that only logical conclusions are valid with certainty, i.e. the direct conclusion resulting from the assumption of $A \rightarrow B$ and A the truth of B concludes. A is in this case a sufficient condition for B while B for A is only necessary. Therefore $A \rightarrow B$ and B are not logically mandatory to conclude A:

Example

1. If the patient has an infection, fever occurs.
2. The patient has a fever.

The patient may have an infection.

Due to other necessary conditions, the clinical picture of an infection may be considered more probable by a physician, but not necessarily true.

Example

A traffic expert has found that

1. When a vehicle leaves the road, the driver has often fallen asleep.
2. The vehicle is off the road.

The driver is probably asleep.

Example

An economic expert states:

1. This is a long-term investment.
2. The desired yield is 10%.
3. The area of the investment is undetermined.

An investment pays off with a certain safety factor.

In the theory of science, statistical reasoning as well as the inductive degrees of confirmation of a hypothesis, which are made dependent on the extent of the confirmations, are extremely interesting [12]. As a basic algorithm for the evaluation of a diagnosis in expert systems, the following procedure is suitable [13]:

1. Start with the assumed ("a priori") probabilities of all (possible) diagnoses;

2. For each symptom, modify the (conditional) probability of all diagnoses (according to the frequency of occurrence of a symptom in the presence of a diagnosis);
3. Select the most likely diagnosis. (Bayes' theorem is often used as a general formula for computing the most likely diagnosis under the assumption of certain symptoms).

Knowledge representations by experts must therefore take uncertainty factors into account. Even the terms used by experts are by no means always clearly defined, and yet they are still used. Information about colour, elasticity and similar qualities only makes sense when referring to certain intervals. The limits of these intervals then appear to be arbitrarily set. Whether for a designer a color is still black or already gray, is considered quite blurred ("fuzzy"). In theory of science, the structure of a "fuzzy logic" is therefore attempted [14]. Paradoxes are inevitable without appropriate interpretation: When a pile of n straws is called large, then there is also a pile of straws with $n - 1$ straws big. If one applies this conclusion iteratedly, the empty heap will consequently also be large.

The representation of knowledge in classical logic is based on the fiction of a temporally unchangeable validity of its conclusions. In fact, however, new information that was not yet included in the knowledge base can invalidate old derivations. Example: If P is a bird, P can fly: Charly is a bird, but also a penguin. Thus, while in classical logic the amount of derivatives increases with the growing amount of assumed facts (monotony), in fact the amount of derivatives can be limited with the growing amount of new information over time (non-monotony). This non-monotony in concluding and judging must also be regarded by an expert as a realistic situation, since a complete and error-free data collection would not be possible, too costly or protracted for an upcoming problem solution.

For an expert system, changing input data of the knowledge base require that the evaluation of conclusions be recalculated. Therefore, the representation of knowledge in databases is now also provided with time specifications. In medical diagnostics, information about the temporal change of a symptom is inevitable. Here, too, scientific theory has done pioneering work with the logic of temporal reasoning, which is now being consciously or unconsciously implemented by the constructors of knowledge-based expert systems [15].

The philosopher H. Dreyfus distinguishes a 5-step model from the beginner to the expert, which should underline this insight [16]. At level 1, the beginner adopts rules that he stubbornly applies without reference to the overall situation. The learner learns how to shift gears at fixed mileage values, the apprentice learns individual parts of an engine, the player learns the basic rules of a game. At level 2, the advanced beginner already occasionally refers to situation-dependent characteristics. The apprentice learns to take into account experience values of certain materials, the learner learns to switch gears due to engine noises and so on and so forth. At level 3, competence has already been achieved and the apprentice has, so to speak, passed the journeyman's examination. The apprentice has learned to design solution strategies for complex problems from the learned rules in his specific area of application. The motorist can properly coordinate and apply the individual rules for driving his vehicle. According to Dreyfus, the maximum performance of an expert system has already been achieved.

The next steps of the master and expert cannot be recorded algorithmically. It requires judgement, which refers to the entire situation, the chess master, who recognizes complex constellation patterns in a flash and compares them with known patterns, the racing driver, who intuitively feels the driving style optimally adapted to the engine and the situation, the engineer, who hears where the engine fault lies on the basis of his experience based on noises.

How do you become a good management expert? Algorithmic thinking, computer-aided problem analysis and the use of expert systems help in the preparation phase. Expert systems lack, as was pointed out above, above all a general world and background knowledge. The sense for the whole as a basis for correct decisions cannot be learned from textbooks or planning calculations. After basic training, a manager no longer learns through abstract definitions and general textbook rules. He learns from his company through concrete examples and cases as far as possible and is able to utilize them in a situation-related way. Concrete case studies combined with a sense for the whole sharpen the judgement of the future manager.

References

1. Puppe LF (1988) Introduction to expert systems. Springer, Berlin
2. Kredel L (1988) Artificial intelligence and expert systems. Droemer Knaur, Munich
3. Mainzer K (1990) Knowledge-based systems. Remarks on the philosophy of technology and artificial intelligence. J Gen Philos Sci 21:47–74
4. Clancey W (1983) The epistemology of a rule-based expert system—a framework for explanation. AI Journal 20:215–293
5. Nilson N (1982) Principles of artificial intelligence. Springer, Berlin
6. Minsky M (1975) A framework for representing knowledge. In: Winston P (ed) The psychology of computer vision. McGraw-Hill, New York
7. Sussmann G, Steele G (1980) Constraints—a language for expressing almost-hierarchical descriptions. AI J 14:1–39
8. Buchanan BG, Sutherland GL, Feigenbaum EA (1969) Heuristic DENDRAL: A program for generating processes in organic chemistry. In: Meltzer B, Michie D (eds) Machine Intelligence, vol 4. Elsevier Science Publishing Co, Edinburgh
9. Buchanan BG, Feigenbaum EA (1978) DENDRAL and META-DENDRAL: Their applications dimensions. Artif Intell 11:5–24
10. Shortliffe EH (1976) Computer-Based medical consultations: MYCIN. Elsevier Science Ltd, New York
11. Randall D, Buchanan BG, Shortliffe EH (1977) Producing rules as a representation for a knowledge-based consultation program. Artif Intell 8:15–45
12. Carnap R (1959) Inductive logic and probability. Springer, Vienna
13. Lindley DV (1965) Introduction to probability and statistics from a bayesian viewpoint I-II. Cambridge University Press, Cambridge
14. Zadeh LA (1975) Fuzzy sets and their application to cognitive and decision processes. Academic Press, New York
15. de Kleer J (1986) An assumption based TMS. AI J 28:127–162
16. Dreyfus HL, Dreyfus SE (1986) Mind over machine. Free Press, New York

Computers Learn to Speak

5

5.1 ELIZA Recognized Character Patterns

Against the background of knowledge-based systems, Turing's famous question, which moved early AI researchers, can be taken up again: Can these systems "think"? Are they "intelligent"? The analysis shows that knowledge-based expert systems as well as conventional computer programs are based on algorithms. Even the separation of knowledge base and problem solving strategy does not change this, because both components of an expert system must be represented in algorithmic data structures in order to finally become programmable on a computer.

This also applies to the realization of natural language through computers. One example is J. Weizenbaum's ELIZA language program [1]. As a human expert, ELIZA will simulate a psychiatrist talking to a patient. These are rules on how to react to certain sentence patterns of the patient with certain sentence patterns of the "psychiatrist". In general, it is about the recognition or classification of rules with regard to their applicability in situations. In the simplest case, the equality of two symbol structures must be determined, as determined by the EQUAL function in the LISP programming language for symbol lists (see Sect. 3.3). An extension exists if terms and variables are included in the symbolic expressions, e.g.

$$(x\ B\ C)$$

$$(A\ B\ y)$$

© Springer-Verlag GmbH Germany, part of Springer Nature 2020   59
K. Mainzer, *Artificial intelligence – When
do machines take over?*, Technik im Fokus,
https://doi.org/10.1007/978-3-662-59717-0_5

The two terms are equal, since the variable x corresponds to the atom A and the variable y to the atom C. By introducing variables, the structure of a pattern can be recognized by pattern matching. For this purpose, a separate operator MATCH is defined in LISP. The simplest case of the correspondence of a date and a structure is the equality of constants, e.g. the equality of the lists AB and AB:

$$\big(\mathrm{MATCH'(AB)'(AB)}\big) = \mathrm{T}$$

Different generators, which act like variables in logic or numerics, lead to class formation. So with the symbol "?" any object can be adjusted, e.g.

$$\big(\mathrm{MATCH'(?B)'(AB)}\big) = \mathrm{T}$$
$$\big(\mathrm{MATCH'(?B)'((AB)B)}\big) = \mathrm{T}$$

The structure therefore compares each two-element list with B as the second element. The *-generator compares any sequence of elements (including the empty sequence), e.g.

$$\big(\mathrm{MATCH'(*B)'(B)}\big) = \mathrm{T}$$
$$\big(\mathrm{MATCH'(*B)'(ABCB)}\big) = \mathrm{T}$$

The given structure matches all lists with the end element B. This already explains the essential basics for understanding the algorithmic structure that reflects the alleged knowledge of the psychiatrist in ELIZA. The dialogue between patient P and ELIZA can be imagined in this way:

Example
P   MEN ARE ALL ALIKE
E   IN WHAT WAY
P   THEY'RE   ALWAYS   BUGGING   US   ABOUT SOMETHING OR OTHER
E   CAN YOU THINK OF A SPECIFIC EXAMPLE
P   WELL; MY BOYFRIEND MADE ME COME HERE
E   YOUR BOYFRIEND MADE YOU COME HERE

    P   YES; I AM VERY SAD
    E   YOU ARE VERY SAD
    P   SAD AND DEPRESSED
    E   PLEASE GO ON

    ...

The following rules allow the derivation of ELIZA's answers if certain sentence patterns are used by the patient and recognized by ELIZA by means of pattern matching:

```
'(((*AM *B) → (APPEND'(YOU ARE)B))
((*ARE*) → '(IN WHAT WAY))
((*ALWAYS*) → '(CAN   YOU   THINK   OF   A   SPECIFIC
EXAMPLE))
((*MY * B ME * C) → (APPEND '(YOUR)(APPEND B(CONS
'YOU C))))
((*L) → '(PLEASE GO ON)))
```

The second rule says: If the patient's sentence contains ARE, answer with the list '(IN WHAT WAY). In the sentence MEN ARE ALL ALIKE, therefore, the *-operator, the list MEN appears before ARE, and the list ALL ALIKE appears after ARE.

The fourth rule states: If the words MY and ME are separated by a list *B in the patient's record, and the record ends with a list *C, then ELIZA responds, first assemble YOU and the C part together (CONS'YOU C), then apply the B part to it, and finally '(YOUR) to it.

So the dialog with ELIZA is nothing else than the derivation of syntactic symbol lists in our example of the program language LISP. Semantically, the structures are chosen so that they correspond to colloquial entertainment habits. The last rule is a typical embarrassment reaction, as it also occurs in actual conversations: If any symbol list (*L) is not recognized by the expert (so to speak the conversation noise blah, blah, blah, blah, ...), then he makes an intelligent face and says PLEASE GO ON.

We must by no means throw out the child with the bathwater and conclude from the simple algorithmic structure of this conversation that it is a mere conjuring trick to simulate the Turing test. The simple example of ELIZA makes it clear that party conversations as well as questioning human experts are predetermined by basic patterns in which we can only vary to a certain degree. These respective basic patterns are described by a number of expert systems. algorithmically captured—no more and no less. In contrast to the expert system, however, humans cannot be reduced to individual algorithmic structures.

5.2   Automata and Machines Recognize Languages

Computers basically process texts as sequences of symbols of a certain alphabet. Computer programs are texts above the alphabet of a computer keyboard, i.e. the symbols of the keys of a keyboard. These texts are automatically translated in the computer into bit sequences of the machine language, i.e. symbol sequences of an alphabet consisting of the two digits 0 and 1, which stand for alternative technical states of the calculating machine. Through these texts and their translation into technical processes, the physical machine of the computer comes into operation. In the following we will first discuss a general system of formal languages which are understood by different types of automats and machines. The natural languages of us humans, but also the means of communication of other organisms, are considered to be special cases under special circumstances (contexts).

▶ **Definition**
An alphabet $\sum$ is a finite (non-empty) set of symbols (also called characters or letters, depending on the application). Examples are

$\sum_{bool} = \{0, 1\}$ Boolean alphabet of machine language,
$\sum_{lat} = \{a, b, \ldots, z, A, B, \ldots, Z\}$ Latin alphabet of some natural languages,

$\sum_{keyboard}$ consists of $\sum_{lat}$ and the other symbols of a keyboard such as B. !, ′, §, $, ... and the space character (as an empty space between symbols).

A word about $\sum$ is a finite or empty sequence of symbols. The empty word is called ε. The length $|w|$ of a word w denotes the number of symbols of a word (with $|\varepsilon| = 0$ for an empty word, but $|\sqcup| = 1$ for the space of the keyboard). Examples of words are

"010010" on the Boolean alphabet $\sum_{bool}$,
"Here we go!" on the alphabet of the keyboard $\sum_{keyboard}$.

denotes the set of all words on the alphabet $\sum$.

Example: $\sum_{bool}{}^* = \{\varepsilon, 0, 1, 00, 01, 10, 11, 000, \ldots\}$

One language L on an alphabet $\sum$ is a subset of $\sum^*$.

The concatenation of words w and v from $\sum^*$ is combined with wv. Accordingly $L_1 L_2$ is the chaining of languages L_1 and L_2 that are derived from the chained words wv with w from L_1 and v from L_2.

When does an automaton or a machine recognize a language?

▶ An algorithm (i.e. Turing machine or, according to Church's thesis, a computer) recognizes a language L on an alphabet $\sum$, if it can decide for all symbol sequences w from $\sum^*$ whether w is a word from L or not.

We distinguish between automata and machines of different complexity that can recognize languages of different complexity [2]. Finite automata are particularly simple automata with which processes can be described without delay on the basis of limited memory [3]. Examples are telephone circuits, adding, operating coffee machines or controlling lifts. Multiplications cannot be carried out with finite automata, as intermediate calculations with delays in processing are necessary for this purpose. This also applies to the comparison of words, since they can be of any length and can no longer be buffered in a limited memory.

One can clearly imagine a finite automat as shown in the Fig. 5.1. There a stored program, a tape with an input word and a read head are distinguished, which can move on the tape only from left to right. This input tape can be understood as a linear memory for the input. It's divided into fields. Each field serves as a storage unit that can contain a symbol of an alphabet $\sum$.

For speech recognition the work of a finite automaton begins with the input of a word w on the alphabet $\sum$. During the input the finite automaton is in a certain state s_0. Each finite automaton is characterized by a set of accepting states (or end states). In the further processing steps, the symbol sequences and the respective states of the machine change, until finally, after many finite steps, the empty word ε in a state s is reached. When this final state s belongs to the distinguished accepting states of the automaton, then the finite automaton has accepted the word. In the other case, the word w is rejected by the automaton. A finite automaton thus accepts an input word if it is in an accepting state after reading the last letter of the input word.

▶ **Definition**
Language accepted by a finite automaton FA L(FA) consists of the accepted words w from $\sum^*$.

The class $\mathcal{L}$(FA) of all languages accepted by a finite automaton FA is called the class of regular languages

Regular languages are characterized by regular expressions (words), which arise from the symbols of an alphabet through alternative, concatenation and repetition. Consider, for example, the alphabet $\sum = \{a, b, c\}$. An example of a regular language

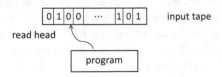

Fig. 5.1 Scheme of a finite automaton

is then the language that contains all words that consist of any number of *a* (repetitions like e.g. *a*, *aa*, *aaa*, …) or (alternative) from any number of *b* (repetitions like *b*, *bb*, *bbb*, …). Another example of a regular language includes all the words that begin with *a* and end with *b* and in between, only repetitions of *c* as for example *acb*, *accccb*.

To show that a language is not regular, it is sufficient to show that there is no finite automaton that accepts it. Finite automata have no other storage possibility than the current state. So if a finite automaton after reading two different words ends in the same state again, it can no longer distinguish between the two words: It "forgot" the difference.

▶ **Definition**
A deterministic finite automaton is determined by deterministic processes. Each configuration is uniquely defined by the state of the machine and the word read. A program completely and unambiguously determines the sequence of configurations from machine states and associated words.

A non-deterministic finite automaton allows a selection of several possible subsequent configurations in certain configurations.

Therefore, a non-deterministic algorithm can be lead to exponentially many possibilities. In general, however, there is no more efficient way to simulate non-deterministic algorithms by deterministic algorithms than to simulate all possible alternatives by a deterministic algorithm. Even in the case of finite automatons, it can be proven that the non-deterministic extension of the possibilities for speech recognition brings nothing new: The deterministic finite automatons accept the same languages as the non-deterministic finite automatons.

A Turing machine (see Sect. 3.4) can be understood as an extension of a finite automaton. It consists of

- a finite control that contains the program,
- an unlimited tape that is used as both input tape and memory,
- a read/write head that can move the tape in both directions.

A Turing machine is similar to a finite automaton in that it works on a finite alphabet using a tape containing an input word at the beginning. Unlike a finite automaton, a Turing machine can also use the unlimited tape as storage. A finite automaton can be extended to a Turing machine by replacing the read head with a read/write head and also moving it to the left [4].

A Turing machine TM is determined by an initial state, an accepting state and a rejecting state. When TM reaches the accepting state, it accepts the input word regardless of where the read/write head is on the tape. When TM reaches the rejecting state, it rejects the input word and stops. A word, however, is rejected by a TM even if it does not stop after its input after a finite number of steps.

▶ Definition
The language accepted by a Turing machine TM L(TM) consists of the accepted words w from $\sum^*$.

The class $\mathcal{L}$(TM) of all languages accepted by a Turing machine TM is called the class of recursively enumerable languages.

A language is called recursive or decidable, if there is a Turing machine TM, which can be used for all words w from $\sum^*$ to decide whether w is accepted (and belongs to the language) or is not accepted (and therefore does not belong to the language).

According to Church's thesis (cf. Sect. 3.4) the Turing machine is the logical-mathematical prototype for a computer at all - independent of its technical realization as a supercomputer, laptop or smartphone. Practical computers, however, have the so-called von Neumann architecture, where the memory for program and data, CPU and input are technically independent units. In a Turing machine, input and memory are merged into one unit of tape, read and write into one read/write head. This is theoretically not a problem, as multi-tape Turing machines which have several tapes with their own read/write head. They then take over the separate functions of the von Neumann architecture.

Logically-mathematically, the one-tape Turing machine is equivalent to the multi-tape Turing machine. That means that the one-tape machine can simulate the multi-tape machine.

Analogous to finite automata, deterministic Turing machines can be extend to non-deterministic Turing machines. A non-deterministic Turing machine can finally follow many alternatives after an input word. These machining operations can be imagined graphically as a branching tree. The input word is accepted if at least one of these operations ends in the accepting state of the Turing machine. A distinction is made between the depth search as the machining strategy for such branching trees from the breadth search. In the depth search, each "branch" of the branching tree is tested one after the other to see whether it ends in an accepted final state. In the breadth search, all branches are tested simultaneously to a certain depth, whether one of them reaches the accepting state. The process is repeated step by step until this happens. Then the machine stops. By a breadth search of the branching tree non-deterministic Turing machines can be simulated by deterministic Turing machines.

In general, no more efficient deterministic simulation of non-deterministic algorithms is known as a step-by-step simulation of all calculations of a non-deterministic algorithm. However, this has its price: when non-determinism is simulated by determinism, the computing time grows exponentially. So far, the existence of a much more efficient simulation is not known. However, the non-existence of such a simulation has not yet been proven.

From natural languages we are used to the fact that their words and sentences are determined by grammatical rules. Each language can be determined by a grammar, i.e. a system of appropriate rules. A distinction is made between terminal symbols such as a, b, c, … and digits of non-terminal symbols (non-terminals) A, B, C, …; X, Y, Z, … Non-terminals are used like variables (spaces), which can be replaced by other words [5].

Example

example of a grammar:
  terminals: a, b
  non-terminals: S
  rules:

$$R_1 : S \rightarrow \varepsilon$$
$$R_2 : S \rightarrow SS$$
$$R_3 : S \rightarrow aSb$$
$$R_4 : S \rightarrow bSa$$

derivation of the word *baabaabb*:

$$S \rightarrow_{R_2} SS \rightarrow_{R_3} SaSb \rightarrow_{R_3} SaSSb \rightarrow_{R_4} bSaaSSb \rightarrow_{R_1} baassb$$
$$\rightarrow_{R_3} baabSaSb \rightarrow_{R_1} baabaSb \cdots \rightarrow_{R_3} baabaaSbb \rightarrow_{R_1} baabaabb$$

Obviously, grammars are non-deterministic methods for the generation of symbol sequences. So several rules with the same left side are allowed. Furthermore, it is not specified which rule is applied first to replacements in a word if several options exist.

In linguistics, grammars are used to syntactically describe natural languages. Syntactic categories such as ⟨*sentence*⟩, ⟨*text*⟩, ⟨*noun*⟩ and ⟨*adjektive*⟩ were introduced as nonterminals. Texts can be derived with appropriate grammar rules.

Example

Text derivation with grammar rules:

$$⟨text⟩ \rightarrow ⟨sentence⟩⟨text⟩$$
$$⟨sentence⟩ \rightarrow ⟨subject⟩⟨verb⟩⟨object⟩$$
$$⟨subject⟩ \rightarrow ⟨adjective⟩⟨noun⟩$$
$$⟨noun⟩ \rightarrow ⟨tree⟩$$
$$⟨adjektive⟩ \rightarrow [green]$$

According to N. Chomsky, a hierarchy of grammars with different complexity can be given [6]. Since the corresponding languages are generated by grammatical rules, he also called them generative grammars:

▶ **Definition**

1. regular grammar:

   The simplest class are the regular grammars, which create exactly the class of regular languages. The rules of a regular grammar have the following form $X \to u$ and $X \to uY$ for a terminal u and the non-terminals X and Y.

2. context-free grammar:

   All rules have the form $X \to \alpha$ with a non-terminal X and a word α from terminals and non-terminals.

3. context sensitive grammar:

   In the rules $\alpha \to \beta$ is the length of word α not greater than the length of word β. Therefore, no partial word α in the derivatives can be replaced by a shorter partial word β.

4. unlimited grammar:

   These rules are not subject to any restrictions.

Context-free grammars differ from regular grammars in that the right side of a regular rule contains at most one non-terminal. Unlike unrestricted grammars, context-sensitive grammars do not contain a rule where the word on the left side is larger than the word on the right side. Therefore, in a computer, the unrestricted grammars can generate arbitrary memory contents and thus simulate arbitrary derivatives.

How do the different grammars relate to machines and machines that recognize these languages? An equivalent finite automaton can be specified for each regular grammar, which recognizes the corresponding regular language. Conversely, an equivalent regular grammar can be specified for each finite automaton, with which the corresponding regular language is generated.

Context-free grammars create context-free languages. A pushdown automaton can be introduced as a suitable automaton type that recognizes context-free languages:

▶ **Definition**

A pushdown automaton (PDA) (cf. Fig. 5.2) has an input tape which contains the input word at the beginning. As with finite automata, the read head can only read and move from left to

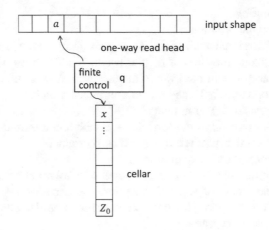

Fig. 5.2 Architecture of a pushdown automaton

right. Therefore, the tape can only be used to read the input and not as a memory like a Turing machine. However, in contrast to a finite automaton, a pushdown automaton does not have to move to the right with the read head after reading a symbol. It can remain on the same field of the tape and make edits on the data in the "cellar". The pushdown automaton can only access and read the top symbol in the cellar. If deeper data is to be accessed, the previous data must be irrevocably deleted. The cellar is in principle an unrestricted tape with finally many access possibilities.

The cellar automat thus begins its work with the read symbol on the input tape, the state of the final control and the top symbol of the cellar. In further actions, it changes the state, moves one field to the right with the read head, and replaces the topmost symbol x of the cellar by one word α.

The non-deterministic pushdown automata recognize exactly the class of context-free languages. The non-deterministic cellar machines are thus equivalent to the context-free grammars, which produce exactly the context-free languages. in computer science, context-free grammars are suitable to represent programming languages. The words generated by context-free grammars correspond to correct programs of the modeled programming language. Context-free grammars are therefore suitable for the construction of compilers. These are computer programs that translate another program written in a particular programming language into a form that can be executed by a computer.

In the Chomsky hierarchy now follow the context-sensitive languages, which are generated by context-sensitive grammars. Context sensitive languages are recognized by a restricted machine type of the Turing machine:

▶ A linearly limited automaton is a Turing machine whose working tape is limited by the length of the input word. Two additional symbols are used which mark the left or right end of the input word and which must not be exceeded during processing.

The set of languages which is recognized by non-deterministic linearly constrained automata is equal to the set of context-sensitive languages. So far it is not proved, whether deterministic linearly limited automata accept the same language class as the non-deterministic ones.

▶ **Important**
The unlimited grammars generate exactly the recursively enumerable languages that can be recognized by Turing machines. The set of recursively enumerable languages is therefore exactly the class of all languages that can be generated by grammars at all.

Languages that cannot be enumerated recursively can thus only be recognized by machines that lie beyond the Turing machine, i.e.—intuitively speaking—"can do more than Turing machines". This is central to the question of AI, whether intelligence can be reduced to Turing machines as prototypes of computers or is more.

Generative Grammars do not only generate syntactic symbol sequences. They also determine the meaning of sentences. Chomsky first analyzed the surface of a sentence as a structure composed of phrases and phrase parts. They were divided into further parts by further rules, until finally the individual words of a sentence of a natural language become derivable. Then, a sentence consists of a nominal phrase and a verbal phrase, a nominal phrase consists of an article and a noun, a verbal phrase consists of a verb and a nominal phrase, etc. Thus, sentences can be characterized by different grammatical depth structures in order to grasp different meanings.

According to this, the same sentence can have different meanings, which are determined by different grammatical depth structures [7]. In Fig. 5.3, the sentence "She drove the man out with the dog" can have the meaning that a woman drove a man out with the help of a dog (a). But, the sentence can also have the meaning that a woman expelled a man who carried a dog with him (b). The production rules are as $\langle S \rangle$ (sentence), $\langle NP \rangle$ (nominal phrase), $\langle VP \rangle$ (verbal phrase), $\langle PP \rangle$ (prepositional phrase), $\langle T \rangle$ (article), $\langle N \rangle$ (noun), $\langle V \rangle$ (verb), $\langle P \rangle$ (preposition), $\langle Pr \rangle$ (pronouns):

$$\langle S \rangle \rightarrow \langle NP \rangle \langle VP \rangle \qquad \langle Pr \rangle \rightarrow [she]$$
$$\langle NP \rangle \rightarrow \langle T \rangle \langle N \rangle \qquad \langle V \rangle \rightarrow [drove]$$
$$\langle NP \rangle \rightarrow \langle Pr \rangle \qquad \langle T \rangle \rightarrow [the]$$
$$\langle NP \rangle \rightarrow \langle NP \rangle \langle PP \rangle \qquad \langle T \rangle \rightarrow [the]$$
$$\langle VP \rangle \rightarrow \langle V \rangle \langle NP \rangle \qquad \langle N \rangle \rightarrow [man]$$
$$\langle VP \rangle \rightarrow \langle VP \rangle \langle PP \rangle \qquad \langle N \rangle \rightarrow [dog]$$
$$\langle PP \rangle \rightarrow \langle P \rangle \langle NP \rangle \qquad \langle P \rangle \rightarrow [with]$$

A generative grammar is a calculation of such recursive production rules, which can also be realized by a Turing machine. With this generative grammar, the two depth structures for the different meanings (a) and (b) are derived (Fig. 5.3).

Natural languages differ only in the surface structure of a sentence. According to Chomsky, the use of production rules is universal. With a Turing program that simulates finitely many recursive production rules, any number of sentences and their depth grammars can be generated.

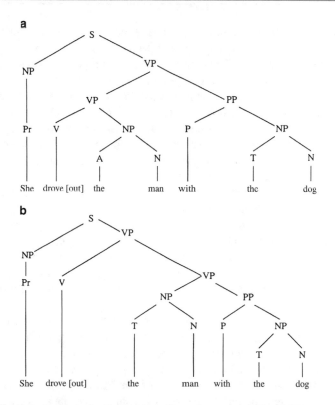

Fig. 5.3 Semantic depth structure a Chomsky grammar [7, p. 347]

The philosopher of language J. Fodor goes beyond Chomsky still because he assumes psychologically real cognitive structures for the linguistic depth structures and universals, which are innate to all human beings [8]. Spirit is understood as a system of semantic representations which are universal and innate and into which all concepts can be decomposed. Fodor speaks of a "language of thought".

However, communication between people is by no means limited to the exchange of ideas about facts. Communication consists of actions of speech that pursue intentions and trigger changes in the environment. The American philosopher J. Searle,

following the British philosopher of language J. L. Austin, introduced the concept of a speech act [9]. A speech act such as e.g.
"Can you give me information about a certain person?" is determined by various action components. First of all, the transmission process of the response must be observed (locutionary
act). The speech act is associated with certain intentions of the
speaker, such as, e.g., request, command or question (illocution-
ary act). The perlocutionary act records the effects of the speech
act on the addressee of a message, e.g. the willingness to give
information about a person or not.

Speech act theory became the model of the computer language KQML (Knowledge and Query Manipulation Language),
which   defines   communication   between   search   programs
("agents") in the Internet. The KQML agent language provides
protocols for mutual identification, establishing a connection
and exchanging messages. At the message level, speech act
types are defined, which can be formulated in different computer
languages.

In technology, the first step is to develop the most efficient
partial solutions possible, which use computer programs to recognize, analyze, transfer, generate and synthesize natural linguistic communication. These technical solutions do not have
to imitate the speech processing of the human brain, but can
also achieve comparable solutions in other ways. Thus, for limited communication purposes of computer programs, it is by no
means necessary that all language layers up to the level of consciousness must be technically simulated.

In fact, in a technically highly developed society, we are also
dependent on implicit and procedural knowledge that can only
be captured in rules to a limited extent. Emotional, social and
situational knowledge in dealing with people can only be formulated with rules to a limited extent. Nevertheless, this knowledge is necessary in order to design user-friendly user interfaces
for technical devices such as computers. Artificial Intelligence
should also be oriented to the needs and intuitions of its users
and not overburden them with complicated rules.

5.3   When Does My Smartphone Understand Me?

Speech comprehension in humans is made possible by the corresponding abilities of the brain. It therefore makes sense to use neural networks and learning algorithms modelled on the brain (see Sect. 7.2). Computational neuroscientist T. J. Sejnowski proposed a neural network that would simulate neural interactions when learning to read in a brain-like machine [10, 11]. Whether the neurons in the human brain actually interact in this way cannot yet be decided physiologically. It remains, however, the amazing achievement that an artificial neural network called NETalk is able to generate a human-like learning process from relatively few neuronal building blocks (Fig. 5.1). The speed of the system could be considerably increased by today's computer power. NETalk also becomes interesting if this system is not simulated on a conventional (sequentially working) computer like all previous artificial neural networks, but could be realized by a corresponding hardware or 'wetware' of living cells.

Example

As input of NETalk, a text is entered character by character (Fig. 5.4; [12]). Since the surrounding characters are important for the pronunciation of a character, the three symbols before and after the character in question are also registered. Each of the seven characters read per step is examined by neurons that correspond to the letters of the alphabet, punctuation, and spaces. The output indicates the phonetic pronunciation of the text. Each output neuron is responsible for a component of sound formation. The conversion of these sound components into an audible sound is performed by an ordinary conventional synthesizer. Decisive is the learning process of reading, which organizes itself between input text and output pronunciation. A third level of neurons is inserted, whose synaptic connections with input and output neurons are simulated by numerical weights.

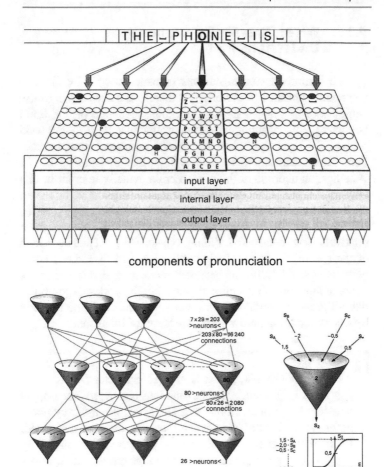

Fig. 5.4 NETalk learns to read [12]

In a training phase, the system first learns the pronunciation of a sample text. No program with explicit rules for sound formation is entered into the system. The pronunciation of the text is rather stored by synaptic interconnection of the neurons. In the case of a foreign text, its initially random pronunciation sounds are compared with the desired sounds of the standard text. If the output is not correct, the system works backwards to the internal level and checks why the networking led to this output, which connections had the highest weight and thus the greatest influence on this output. It then changes the weights to gradually optimize the result. So NETalk works after the learning algorithm of backpropagation which dates back to D. Rumelhart et al. (see Sect. 7.2).

The system learns to read in a similar way to a human being through "learning by doing" and not rule-based. In ever new reading attempts the system improves like a schoolchild its pronunciation and has finally an error rate of approx. 5 %.

But, do we really first need knowledge of neural speech processing in the brain in order to use AI software for speech processing? With the growing performance of computers, individual works such as e.g. by Galilei and Thomas von Aquinas have already been digitally stored and catalogued in the past. Google opened up new possibilities for processing the systematic digitization of literature worldwide, which is now called "digital humanities" [13, 14]. Actually, the methods of digital humanities go beyond the mere digitization of texts and make use of Big Data methods (see Sect. 10.1). An essential approach of Big Data is that you do not have to know the content in detail to derive certain information from data. In the research field of eCodicology, metadata of old manuscripts are created algorithmically in order to draw conclusions about their places of origin, conditions of production, and contextual relations. Meta data concern, e.g., page format, inscriptions, registers, or marginal notes.

The ePoetics project investigates the spread of literary terminology in a historical period. From this, conclusions can be drawn about the development of literary theory during this period. A single scientist can read only a limited number of texts. In order to capture and categorize epochs and styles, thousands of novels and short stories may be necessary. Appropriate software can quickly deliver correlations and illustrate them vividly in diagrams. However, there is a critical reservation: in the end, the supercomputer does not replace the literary scholar's evaluation and interpretation. However, as the semantic web shows, suitable software is able to recognize semantic contexts. Literary scholars who still believe that computers "only" syntactically change symbols have not yet understood the seriousness of the situation and their subject.

The next step is to use software agents (bots) that automatically compose texts. With simple texts, as they are usual in the social media, this will come as no surprise. Are we already tweeting with bots instead of people? But, even in certain areas of journalism, bots replace or at least support copywriters. The company Narrative Science offers software to automatically create articles in journals. Companies use these writing programs for, e.g., automated stock exchange reports. The writing programs can adapt in style to an author. By connecting to a database, the text can be published quickly. Banks use the texts and can react immediately to new data in order to make profits faster than their competitors. Again, it is remarkable and typical for Big Data that it is not the correctness of the data that counts, but the speed of reaction. As long as all parties use the same data, the quality and reliability of the information will not affect the chances of winning.

Text comparisons on the basis of pattern recognition have been known since the introduction of ELIZA. Today's software now breaks sentences down into individual phrases and calculates the probabilities for suitable answer patterns to questions asked or suitable translations into other languages at speed of light. An example of an efficient translation program has already been VERBMOBIL.

Example

VERBMOBIL was a project coordinated in 1993–2000 by the German Research Center for Artificial Intelligence (DFKI) [15]. In detail, the spoken language was transmitted via two microphones to the speech recognition modules for German, English or Japanese and subjected to a prosody analysis (analysis of the speech metrics and rhythm). On this basis, meaning information was taken into account in an integrated processing, which was obtained by grammatical depth analyses of sentences and rules of dialogue processing. VERBMOBIL thus realized the transition from the recognition of colloquial speech to the dialogue semantics of conversations, which were by no means limited to the exchange of short language lumps, but also included long speeches, as they are typical for spontaneous language.

Speech processing goes through different levels of representation with us humans. In technical systems one tries to realize these steps one after the other. In computer linguistics [16–18] this procedure is described as a pipeline model:

Starting from sound information (hearing), a text form is generated in the next step. The corresponding strings of letters are then recorded as words and sentences. In the morphological analysis, personal forms are analyzed and words in the text are traced back to basic forms. In the syntactic analysis, the grammatical forms of the sentences such as subject, predicate, object, adjective etc. are highlighted, as explained in the Chomsky grammars (cf. Sect. 5.2). In the semantic analysis, meanings are assigned to the sentences, as was done in the depth structures of the Chomsky grammars. Finally, in a dialogue and discourse analysis, the relationships between e.g. question and answer, but also intentions, purposes and intentions are examined.

As we will see later, it is by no means necessary for efficient technical solutions to pass through all stages of this pipeline model. Today's enormous computing power, combined with machine learning and search algorithms, open up the exploitation of data patterns that can be used for efficient solutions at all levels. Generative grammars for the semantic analysis of depth structures are hardly used for this purpose. Also the orientation at the semantic information processing of humans does not play a role. In humans, semantic processes are typically associated with consciousness, which is by no means necessary:

Example

A semantic question-answer system is the program WATSON from IBM, which uses the computing power of a parallel

computer and the memory of Wikipedia. In contrast to
ELIZA, WATSON understands the semantic meanings of
contexts and language games. WATSON is a semantic search
engine (IBM) that captures questions posed in natural language and finds suitable facts and answers in a large database
in a short time. It integrates many parallel language algorithms, expert systems, search engines and linguistic processors on the basis of the computing and storage capacities of
huge amounts of data (big data) (Fig. 5.5; [19]).

WATSON is not oriented towards the human brain, but relies on
computing power and database capacities. Nevertheless, the system passes the Turing test. A style analysis adapts to the habits
of the speaker or writer. Personalization of the writing style is
therefore no longer an insurmountable barrier.

WATSON now refers to an IBM platform for cognitive tools
and their diverse application in business and enterprises [20].
According to Moore's law (see Sect. 9.4), WATSON's services
will not require a supercomputer in the foreseeable future. Then
an app in a smartphone will deliver the same performance. We'll
finally be talking on our smartphone. Services no longer have to
be requested via a keyboard, but by speaking with an intelligent
speech program. Even conversations about our intimate feelings
cannot be ruled out, as Weizenbaum had feared.

The US science fiction film "Her" by Spike Jonze from 2013 is about
an introverted and shy man who falls in love with a language program.
Professionally, this man writes letters on commission for people who find
it difficult to communicate their feelings to their counterparts. For his own
relief, he procures a new operating system equipped with a female identity
and a pleasant voice. Via headset and video camera, he communicates with
Samantha how the system calls itself. Samantha quickly learns about social
interactions and behaves more and more humanely. During frequent, long and
intense conversations, an intimate emotional relationship eventually develops.

The use of intelligent writing programs is not only conceivable
in media and journalism when it comes to routine texts from e.g.
business news, sports reports or tabloid announcements. Routine
texts that can be delegated to bots are also used in administration

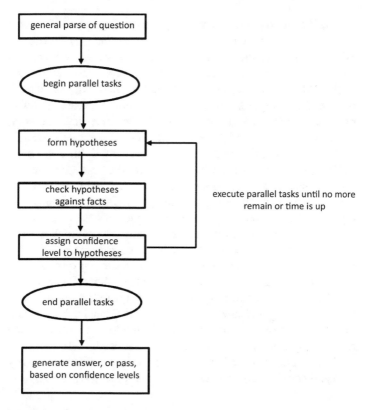

Fig. 5.5 Architecture of WATSON [19]

or jurisdiction. We will also experience the use of automatic writing programs in science. The production of articles in medical, technical and natural science journals has become so gigantic that even in special areas of research, the respective experts can no longer see them in detail. The research results must be published with great speed in order to survive in competition. So it is quite conceivable that scientists and scholars only enter data, arguments and results in the usual linguistic structure (e.g. of a preprint), which a bot adapts to the writing style of the author and publishes via a database.

Writing robots are increasingly becoming part of every-day life in the financial sector. Companies such as "Narrative Science" or "Automated Insight" use intelligent software to translate quarterly data from investment banks into news texts. Such texts were previously written by reporters in laborious quarterly reports. Automatic machines generate a multiple of the reports previously written by humans in a matter of seconds. In the financial sector, algorithms generate company profiles for analysis departments at speed of light. Automatic writing programs can inform clients about the strategies used by fund managers to invest money in the stock market and how the funds are performing. Insurance companies use intelligent writing programs to measure sales performance and explain recommendations for improvement. Automatically generated texts enable customers to confirm whether their investment strategy is correct. Support through automated writing programs also creates more time for individual customer advice. With Robo-Advice, artificial intelligence is increasingly advancing in investment consulting and asset management. If German, French and Spanish are now used as languages in addition to English, the area of application will increase. The human investment advisor is not being replaced, but the pace of digital offerings is very high and coordinated with the exponential growth of IT tools.

References

1. Weizenbaum J (1965) ELIZA—A computer program for the study of natural language communication between man and machine. Commun Assoc Comput Mach 9:36–45
2. Hotz GH, Walter H (1968–1969) Automatentheorie und formale Sprachen I–II. B.I. Wissenschaftsverlag, Mannheim
3. Böhling KH, Indermark K, Schütt D (1969) Endliche Automaten I-II. B.I. Wissenschaftsverlag, Mannheim
4. Hopcroft JE, Motwani R, Ullman J (1969) Introduction to automata theory, languages, and computation. Addison Wesley, Readings
5. Hromkovic J (2011) Theoretische Informatik. Formale Sprachen, Berechenbarkeit, Komplexitätstheorie, Algorithmik, Kommunikation und Kryptographie, 4th edn. Vieweg Teubner, Wiesbaden
6. Chomsky N (1969) Aspekte der Syntax-Theorie. Suhrkamp, Frankfurt

7. Anderson JR (1996) Kognitive Psychologie, 2nd edn. Spektrum Akademischer Verlag, Heidelberg
8. Fodor JA (1975) The language of thought. Harvard University Press, New York
9. Searle JR (1969) Speech acts. Cambridge University Press, Cambridge (Mass.)
10. Sejnowski TJ, Rosenberg CR (1986) NETalk: a parallel network that learns to read aloud. In: The John Hopkins University Electrical Engineering and Computer Science Technical Report JHU/EECS-86/01
11. Sejnowski TJ, Rosenberg CR (1987) Parallel networks that learn to pronounce English text. Complex Systems 1:145–168
12. Kinzel W, Deker U (1988) Der ganz andere Computer: Denken nach Menschenart. Bild der Wissenschaft 1:43
13. Schreibman S, Siemens R, Unsworth J (2004) A companion to digital humanities. Wiley-Blackwell, Oxford
14. Thaler M (2012) Controversies around digital humanities. Hist Res 37(3):7–229
15. Wahlster W (ed) (2000) Verbmobil: foundations of Speech-to-Speech Translation. Springer, Berlin
16. Hausser R (2014) Foundations of computational linguistics: human-computer communication in natural language, 3rd edn. Springer, Berlin
17. Jurasky D, Martin JH (2008) Speech and language processing. An introduction to natural language processing, computational linguistics and speech recognition, 2nd edn. PEL, Upper Saddle River
18. Mitkov R (ed) (2003) The oxford handbook of computational linguistics. Oxford University Press, Oxford
19. Watson. (Künstliche Intelligenz) Wikipedia. https://en.wikipedia.org/wiki/Watson_(computer). Accessed 30 Aug 2015
20. http://www-05.ibm.com/de/watson/. Accessed: 30 Aug 2015

Algorithms Simulate Evolution

6

6.1 Biological and Technical Circuit Diagrams

Information processing with computers and humans is reproduced with artificial or natural languages. They are only special cases of symbolic representation systems, which can also be specified for genetic information systems. Genetic languages with their grammatical rules represent molecular methods to generate molecular sequences with genetic meanings. The key to understanding these molecular languages is not us humans, but the molecular systems that make use of them. We humans with our kind of information processing are only at the beginning to decipher and understand these languages with their rules. The formal language and grammar theories, together with the algorithmic complexity theory, provide the first approaches.

For genetic information, the nucleic acid language with the alphabet of the four nucleotides and the amino acid language with the alphabet of the twenty amino acids are used. In the nucleic acid language, a hierarchy of different language layers can be distinguished, starting at the lowest level of the nucleotides with the basic symbols A, C, G, T or U to the highest level of the genes in which the complete genetic information of a cell is stored. Each intermediate language level consists of units of the previous language level and gives instructions for various functions such as, e.g., transcription or replication of sequences.

© Springer-Verlag GmbH Germany, part of Springer Nature 2020
K. Mainzer, *Artificial intelligence – When do machines take over?*, Technik im Fokus,
https://doi.org/10.1007/978-3-662-59717-0_6

Chomsky's formal language theory can also be used to define the grammar of genetic languages [1, 2].

▶ A. Lindenmayer distinguishes between the following finite alphabets for DNA four letters A, C, G, T, for the RNA four letters A, C, G, U, and for the proteins twenty letters A, C, D, ..., Y, W. The sequences that can be formed from the letters of these alphabets are called words. The set of all sequences forms a language. A grammar consists of rules that convert sequences into others.

Example

In the simplest case of a regular grammar with rules like e.g. $A \rightarrow C, C \rightarrow G, G \rightarrow T, T \rightarrow A$ sequences such as GTACGTA ...can be generated: Start with the left letter of the sequence and add to the right the letters according to the preceding letters.

Regular Grammars determine regular languages which are generated by finite automata as corresponding information systems (see Sect. 5.2). A finite deterministic automaton can be imagined as a machine with finite inputs, outputs and a finite memory that receives, processes, and passes on information.

Finite automata with regular languages are not sufficient to create all possible combinations, like e.g. mirror symmetric sequences AGGA show. For non-regular languages, Chomsky distinguishes a hierarchy of stronger information systems. For this purpose, the grammatical rules of the languages are loosened and the automata are supplemented by memory cells. An example is the pushdown automaton with context-free languages which are dependent on left and right neighbors for the permissibility of a symbol. If this provision is removed, context-dependent languages are obtained in which symbols that are far apart are related to each other. Such context-sensitive languages are recognized by linearly limited automata, in which each of finitely many memory cells can be accessed optionally.

▶ In regular, context-free, and context-sensitive languages, you can decide recursively whether a character sequence of finite length belongs to the language or not. All you need to do is create all character strings up to this length and compare them with the existing character sequence. Genetic languages are of this kind.

If this requirement is not met, machines will be necessary of the complexity the Turing machine (Fig. 3.2). A Turing machine is a finite automaton, so to speak which has free access in an unlimited amount of memory. From this point of view, a linearly limited automaton is a Turing machine with a finite storage tape. The pushdown automaton has a tape that is infinitely long on one side, with the reading head always above the last labelled tape. A finite automaton is a Turing machine without tape. However, there are such complex non-recursive languages whose character series even a Turing machine cannot recognize in finite time. However, they can be mastered intellectually by humans and must therefore be taken into account for the assessment of human information processing within the framework of AI research.

▶ The hierarchy of speech recognition corresponds to different degrees of complexity of the problem solution, which can be mastered by appropriate automata and machines. According to our definition of artificial intelligence (see Chap. 1), these automata and machines have different degrees of intelligence. Therefore, degrees of intelligence can even be assigned to biological organisms, which are examples of these automata and machines.

Genetic languages and their grammars thus permit the determination of the complexity of corresponding genetic information systems. For this purpose, the shortest possible generative grammars are used to generate the respective DNA sequence. The complexity of a genetic grammar is defined by the sum of the length of all rules. Remarkable is also the genetic redundancy of genetic information.

▶ **Definition**
In information theory, redundancy refers to superfluous information. In telecommunications, redundancy is used, for example, by repetitions to protect against transmission errors. In colloquial language, people use repetitions through similar paraphrases to increase understanding.

If there is no redundancy, the average information content (information entropy) of a sequence is maximal. Redundancy measures the deviation of the information content of a sequence from the maximum information entropy.

Measurements show a low redundancy of the genetic languages compared to the human colloquial languages. On the one hand, this shows the great reliability of genetic information systems. On the other hand, redundancy also expresses flexibility and sensitivity to achieve understanding through varied repetitions. It therefore distinguishes human information processing and will have to be taken into account in the assessment of artificial intelligence.

In the evolution of genetic information systems, we observe a tendency to store ever larger amounts of information in order to safeguard the structure and function of an organism. However, it is not only a question of deciphering a DNA sequence as long as possible. In more highly developed organisms such as humans, proteins are of fundamental importance as carriers of a variety of hereditary predispositions. The evolution of these systems is based on the introduction of new rules into the grammar of proteins and thus new functions of the organism. It is even possible to show that the rule length of the protein structures and hence the complexity of their grammar for the more highly developed organisms increases. Grammar rules also allow sequences to be reproduced as often as desired. They therefore contribute to the standardization of genetic information systems, without which the growth of complex organisms would not be possible.

In the evolution of life, the ability to process information is by no means limited to genetic information systems alone. Information processing with nervous systems and brains has paramount importance in highly developed organisms and finally with communication and information systems in populations and societies of organisms [3, 4]. Genetic information systems were formed about three to four billion years ago and have led to a large variety of cellular organisms. This resulted in information storage that can be estimated from the storage capacity of the genetic information generated during evolution.

▶ **Definition**
Generally, the information capacity of a memory is measured by the logarithm of the number of different possible states of the memory. For nucleotide sequences of length n which are formed from four building blocks, there are $4n$ various options of arrangement. Converted to bit units in the dual system, the information capacity is as follows $I_k = \ln 4^n / \ln 2 = 2n$.

For polypeptides from twenty different building blocks, this results in a corresponding storage capacity of $I_k = \ln 20^n / \ln 2 = n \cdot 4,3219$ bit.

For chromosomes with approx. 10^9 nucleotides follows a storage capacity of double length with approx. $2 \cdot 10^9$ bit.

Information and storage capacity are defined independently of the material form of a memory and therefore allow a comparison of different information systems. For comparison, the information capacity of a human storage system such as, e.g., books and libraries can be used:

Example
For one of the 32 letters of the Latin alphabet, $\ln 32/\ln 2 = 5$ bit are required. Therefore, with a DNA sequence, $2 \cdot 10^9$ bit/5 bit $= 4 \cdot 10^8$ letters can be stored. For an average word length of 6 letters, this is approx. $6 \cdot 10^7$ words. A printed page with approx. 300 words results in $2 \cdot 10^5$ printed pages. With a book volume of 500 pages, a DNA sequence consisting of 10^9 nucleotides corresponds to a storage capacity of 400 books.

Today, it is estimated that the biological evolution on earth about ten million years ago after the emergence of bacteria, algae, reptiles and mammals reached a climax of 10^{10} bit of genetic information with the human species.

With the development of nervous systems and brains, new information systems emerged in evolution. However, they were by no means there all at once, but developed through the specialization of some cells in signal transmission. Accordingly, the amounts of information that could be stored by early nervous systems were initially much smaller than in genetic information systems. Only with the appearance of organisms with the complexity of reptiles do neuronal information systems begin to exceed the information capacity of genetic information systems (Fig. 6.1).

So, increased flexibility and the ability to learn are combined in the confrontation with the environment of an organism. In a genetic system, not all possible situations of a complex and constantly changing environment can be considered in program lines of genetic information. In complex cellular organisms, genetic information systems come up against limits and must be supplemented by neuronal information systems [5].

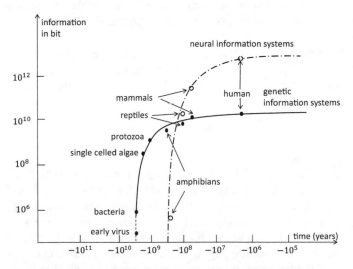

Fig. 6.1 Evolution of genetic and neuronal information systems [3]

6.2   Cellular Automata

The discovery of the DNA structure and genetic code was a first step towards understanding a molecular mechanism that reproduces itself. The computer pioneers John von Neumann and Konrad Zuse showed independently of each other that for self-reproduction not the nature of the material building blocks is fundamental to self-reproduction, but an organizational structure that contains a complete description of itself and uses this information to create new copies (clones) [6, 7].

The analogy with the cell structure of an organism arises when one imagines the system of a cellular automaton as an unlimited chessboard on which each square represents a cell. The individual cells of the parquet plane can be used as finite automata whose finitely many states are distinguished by different colors or numbers. In the simplest case, there are only the two states "black" (1) or "white" (0). An environmental function indicates which other cells the individual cell is connected to. It can define the shape of a cross or square for example.

The state of a cell depends on states in the respective environment and is determined by (local) rules. Since all rules are executed in one step, the network of the cellular automat works synchronously and clockwise. The configuration resulting from a configuration of cellular states by rule application is called successor of the original configuration. The configurations resulting from a configuration by repeated rule application are called generations of the original configuration. A configuration is stable if it matches its successor. It "dies" in the next generation when all its cells are in the state "white" (0).

An example for a 2-dimensional cellular automaton with two states "alive" (black) and "dead" (white) is the following version of J. Conway's "Game of Life" with local rules: (1) A living cell survives to the next generation when two or three cells of the neighborhood are also alive. (2) A cell dies if there are more than three ("overpopulation") or less than two living cells in the neighbourhood ("isolation"). (3) A dead cell may only come alive if exactly three of the neighbouring cells are alive. Conway's "Game of Life" is a cellular automaton which generates complex patterns in subsequent generations, reminiscent of the shape of cellular organisms. It is even a universal cellular automaton, since it can simulate any pattern formation of a cellular automaton [8].

Technically, cellular automata can be simulated by a computer. A corresponding computer program for a cellular automaton uses in principle the same methods, as one would carry out cellular pattern development with paper and pencil. First, a work area is defined for the cells. Each cell corresponds to a memory element in the computer. At each development step, the program must search for each cell one after the other, determine the states of the neighboring cells and calculate the next state of the cell. In this case a cellular automaton can be simulated on a sequential digital computer. Better and more effective would be a network of many processors in cellular interconnection in which the processing takes place in parallel as in a cellular organism. Conversely, every computer can be used as a universal Turing machine with a universal cellular automaton.

According to John von Neumann, a self reproducing automaton must have the performance of a universal Turing machine, i.e. be able to simulate any kind of cellular automaton. In prebiological evolution, the first self-reproducing macromolecules and microorganisms certainly did not have the degree of complexity of a universal computer. Therefore, C. Langton developed (1986) some simpler cellular automata without the ability of universal computability, which can spontaneously reproduce in certain periods like organisms. Its PC images are vividly reminiscent of simple cellular organisms with small tails from which similar small organisms are formed:

Example

The states of the cells are marked in Fig. 6.2 by numbers [9, 10]. Empty cells have the state 8 and form the virtual environment of virtual organisms. Cells in state 2 envelop the virtual organism like a skin and separate it from the environment.

The inner loop carries the code for self-reproduction. At any time, the code numbers are moved counterclockwise step by step. Depending on which code number reaches the tail-like end, it is extended by one unit or a left bend is caused.

After four runs, the second loop is completed. Both loops separate, and the cellular automaton has reproduced itself. Finally, a colony of such organisms covers the screen. While they reproduce at the outer edges, the middle ones are blocked in self-production by their own offspring. Like a coral reef, they form a dead cellular skeleton on which virtual life evolves.

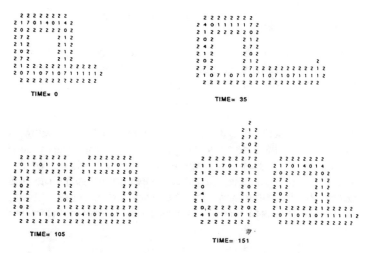

Fig. 6.2 Cellular automata simulate cellular self-organization [9, 10]

Two-dimensional cellular automata consist of a network of individual cells characterized by the geometry of the cell arrangement, the neighborhood of each individual cell, its possible states, and the transformation rules for future states that depend on them. One-dimensional cellular automata consisting of a row of cells in a two-dimensional parquet floor [11, 12] are sufficient for the analysis of evolutionary models. In a simple case each cell has two states 0 and 1, which can be represented graphically by a white or black square. The state of each cell changes in a sequence of discrete time steps according to a transformation rule in which the previous states of the respective cell and its two neighboring cells are taken into account.

▶ **Definition**
In general, the environment function fixes $2r + 1$ cells, i.e. for $r = 1$, in the simplest case, three cells with one preceding cell and two neighbouring cells. Depending on the number of states and neighborhood cells, there are simple local rules with which the discrete temporal development is determined row by row. For $r = 1$ and $k = 2$, it results in $2^3 = 8$ possible distributions of the states 0 and 1 to $2 \cdot 1 + 1 = 3$ cells, i.e. for example the rules:

$$\underline{111} \quad \underline{110} \quad \underline{101} \quad \underline{100} \quad \underline{011} \quad \underline{010} \quad \underline{001} \quad \underline{000}$$
$$0 \qquad 1 \qquad 0 \qquad 1 \qquad 1 \qquad 0 \qquad 1 \qquad 0$$

An automation with these rules has the binary code number 01011010 or (in decimal coding) $0 \cdot 2^7 + 1 \cdot 2^6 + 0 \cdot 2^5 + 1 \cdot 2^4 + 1 \cdot 2^3 + 0 \cdot 2^2 + 1 \cdot 2^1 + 0 \cdot 2^0 = 90$. For 8-digit binary code numbers with two states, there are $2^8 = 256$ possible cellular automata.

With simple local rules, the 256 one-dimensional cellular automata with two states and two neighboring cells can already generate differently complex patterns that are reminiscent of the structures and processes of nature. Their initial states (i.e. the pattern of the initial line) may be ordered or unordered. From this, these automata develop typical final patterns in successive lines.

Fig. 6.3 Cell as dynamic system with state variable x_i, output variable y_i, and three constant binary inputs u_{i-1}, u_i, u_{i+1}

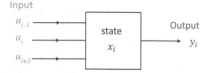

Modern high-speed computers with high storage capacity allow computer experiments to study different patterns of cellular automata. Some automata produce colorful symmetries reminiscent of the patterns of animal skins or shellfish. Others reproduce oscillating wave patterns. After a few steps, some automata develop into a constant state of equilibrium like a molecular system that solidifies in a crystal. Still other automata are sensitive to even the smallest changes of their initial states, which are building up to global changes of their pattern formation. This is reminiscent of currents and patterns from weather and climate models.

In fact, cellular automata like dynamic systems in physics can be described by mathematical equations. The artificial life that they simulate by their pattern formation can then be precisely explained and predicted by solving these equations [13].

▶ In two-dimensional cellular automata, the state x_i of a cell i ($1 \leq i \leq I$) from its own input u_i and the binary inputs u_{i-1} and u_{i+1} of the left and right neighbouring cells (Fig. 6.3). Therefore, the dynamics of the cellular automaton, i.e. the temporal state evolution $\dot{x}_i$ of its cells by equations of state $\dot{x}_i = f(x_i; u_{i-1}, u_i, u_{i+1})$ with an initial condition $x_i(0) = 0$ and output equation $y_i = y(x_i)$ is determined.

6.3   Genetic and Evolutionary Algorithms

Darwin's evolution can be understood as a search for successful species that have adapted to new environmental conditions over generations. Genetic algorithms optimize populations of chromosomes in subsequent generations by reproduction, mutation, and selection [14, 15]. Instead of populations of chromosomes as information carriers, sets of documents would have to be carried out in information retrieval in the Internet [16, 17]. In general, genetic algorithms are applied to sets of binary sequences encoding information strands. Mutation means the random change of a bit from e.g. 01 0 1011 to 01 1 1011. Similar to genetic algorithms where gene sequences can be cut up and reassembled by enzymes, genetic algorithms use recombination methods such as, e.g., crossover, in which the cut sections of two binary sequences are reassembled in reverse:

Selection means the choosing of chromosomes which, after evaluation of the population of a generation, will provide the maximum level of fitness. In the end, the species with the highest levels of fitness should "survive" (survival of the fittest). In simple terms, such an algorithm can be written down in such a way that the subsequent generations of a population P by the numbering i can be counted:

```
i = 0
InitializePopulation P(i);
Evaluate P(i);
while not done {
i = i+1;
P'=SelectParents P(i);
Recombine P'(i);
Mutate P'(i);
Evaluate P'(i)
P=Survive P, P'(i);
}
```

"Survival of the Fittest" can consist of classification tasks for example:

Example

An automaton must decide whether more than 50% of the cells of a randomly selected initial condition are in state 1 ("black") or not:

If this is the case, the automaton will strive for a state of equilibrium in which all cells are in state 1.

In the other case, he strives for a state of equilibrium in which all cells are in the state 0.

The evolution of a population of automata means that genetic algorithms with mutation, recombination, and selection lead to the optimization of machine generations. In the example, for a machine type with $r = 3$ and $k = 2$, there are total 2^7 possible distributions of the states 0 and 1 on $2 \cdot 3 + 1 = 7$ cells, i.e. 128 rules per automaton and 2^{128} automata.

This large class of automata requires a genetic algorithm of optimization in order to solve the above mentioned classification task. A graphic representation of their evolution initially shows a strong improvement in fitness levels, which finally change into saturation in the 18th generation.

Genetic algorithms are not only used to simulate evolutionary processes in the computer. We can also learn from nature and use it to solve problems [18, 19]. Genetic algorithms are used to find optimal computer programs that solve certain problems. Therefore, the program is not explicitly written by the programmer, but generated in the evolutionary process. As in nature, however, there is no guarantee of success. In this case, the virtual organisms are represented by computer programs. Operations of a genetic algorithm optimize generations of computer programs, including a successful specimen.

In nature, a population does not develop under constant environmental conditions. In fact, many evolutions take place simultaneously, with the changing populations interacting as an environment. In biology. in this case, one speaks of co-evolution. They are an example of parallel problem solving and information processing in nature. In this case, fitness levels can be assigned not only to subsequent generations of a population, but also relative to corresponding generations of another population in co-evolution. Practically, several program developments could be tested in such a way by, e.g., competing companies.

The randomness of genetic algorithms does not always meet with approval from programmers. Genetic algorithms work in a similar way to natural selection in evolution: many approaches are tried simultaneously, most of which do not work, few of which can, but do not have to lead to success. Sometimes small DNA flaws cause terrible diseases. On the other hand, our DNA is flawed. These mistakes might then synthesize almost the same amino acids of a protein as the correct code without harming us. Learning from nature therefore also means learning to deal with mistakes. In any case, errors in the software of virtual life are more tolerable than in the wetware of nature:

▶   Learning from nature means learning to program.

Instead of populations of cellular automata, we can also imagine populations of mobile agents developing in a virtual evolution in the computer network. Generations of mobile agents, who look for information according to the specifications of human users, improve their fitness: For the user, they train, e.g., interesting sample articles with examples and look for similar articles in the net. Mobile agents that frequently return from the network with irrelevant information material are selected. Successful agents multiply through mutations and combinations of their charac-teristics. In this case, we are talking about Artificial Life (AL) agents. Populations of AL agents can evolve over generations using genetic algorithms to optimize their online information search for a user's search queries.

In a semantic network, the documents are automatically sup-plemented with meanings, definitions, and concluding rules. If a document is relevant to a user, then this also applies to the links in this document to its supplements, which contribute to increas-ing understanding. In particular, links near keywords in the document are more meaningful than other links. AL agents can make a significant contribution to reducing the search space [20].

Example

First, agents are initialized (initialize agents), receive documents from the user and start with a user profile. In a simplified model, they evaluate the relevance of a document by determining the distance (number of links) between its keywords and the keywords of the query.

The genotype of a AL agent is determined by the variables "confidence" and "energy".

Trust is the degree to which an agent relies in the descriptions of a document and the associated links.

The "Life Energy" E_a of a AL agent a grows or falls with its search success of documents D_a near the search query.

If this energy is greater than a critical value ε, parents bring offspring into the world whose genotype is altered by mutation.

If the power is less than the critical value, the agent dies.

The user profile always has to be adapted again and again in each agent generation (update user profile).

In short: Successful search agents are selected, mutate their genotype, and are allowed to reproduce:

```
Initialize agents;
Obtain queries from user;
while (there is an alive agent){
Get document D_a pointed by current agent;
Pick an agent a randomly;
Select a link and fetch select documer D_a';
Compute the relevancy of document D_a';
Update energy (E_a) according to the document
relevancy;
if (E_a > ε)
Set parent and offspring's genotype appropriately;
Mutate offspring's genotype;
else if (E_a < 0)
Kill agent a;
}
Update user profile;
```

References

1. Lindenmayer R, Rozenberg G (eds) (1976) Automata, languages, development. North-Holland, Amsterdam
2. Lenneberg EH (ed) (1972) Biologische Grundlagen der Sprache. Suhrkamp, Frankfurt
3. Goonatilake S (1991) The evolution of information. Pinter Publishers, London
4. Haefner K (ed) (1992) Evolution of information processing systems. An interdisciplinary approach for a new understanding of nature and society. Springer, Berlin
5. Sagan C (1978) Die Drachen von Eden. Das Wunder der menschlichen Intelligenz. Droemersche Verlagsanstalt Th. Knaur Nachf, München
6. von Neumann J (1966) Theory of self-reproducing automata. University of Illinois Press, Urbana
7. Zuse K (1969) Rechnender Raum. Vieweg + Teubner, Braunschweig
8. Berlekamp E, Conway J, Guy R (1982) Winning ways, vol 2. A K Peters/CRC Press, New York
9. Langton CG (ed) (1989) Artificial life. Westview, Rewood City
10. Langton CG (ed) (1991) Artificial Life II. Westview, Redwood City
11. Wolfram S (1986) Theory and applications of cellular automata. World Scientific Pub Co Inc, Singapur
12. Wolfram S (2002) A new kind of science. Wolfram Media, Champaign/Ill
13. Mainzer K, Chua L (2011) The universe as automaton. From simplicity and symmetry to complexity. Springer, Berlin
14. Holland J (1975) Adaption in natural and artificial systems. A Bradford Book, Ann Arbor
15. Rechenberg I (1973) Evolutionsstrategie: Optimierung technischer Systeme nach Prinzipien der biologischen Evolution. Frommann-Holzboog, Stuttgart
16. Kraft DH, Petry FE, Buckles BP, Sadavisan T (1997) Genetic algorithms for query optimization in information retrieval: relevance feedback. In: Sanchez E, Zadeh LA, Shibata T (eds) Genetic algorithms and fuzzy logic systems. Soft computing perspectives. World Scientific Pub Co Inc, Singapur, pp 155–173
17. Goldberg DE (1989) Genetic algorithms in search, optimization, and machine learning. Addison Wesley, Reading (Mass)
18. Koza JR (1994) Genetic programming II: Automatic discovery of reusable programs. MIT Press, Cambridge
19. Fogel DB (1995) Evolutionary computation: towards a new philosophy of machine intelligence. A Bradford Book, Piscataway N.J.
20. Cho S-B (2000) Artificial life technology for adaptive information processing. In: Kasabov N (ed) Future directions for intelligent systems and information sciences. The future of speech and image technologies, brain computers, www, and bioinformatics. Physica, Heidelberg, pp 13–33

Neuronal Networks Simulate Brains

7

7.1 Brain and Cognition

Brains are examples of complex information systems based on neuronal information processing [1]. What distinguishes them from other information systems is their ability to cognition, emotion and consciousness. The term cognition (lat. cognoscere for "to recognize", "to perceive", "to know") is used to describe abilities such as perception, learning, thinking, memory and language. Which synaptic signal processing processes underlie these processes? Which neuronal subsystems are involved?

In evolution, only a few examples of such cognitive information systems were trained under certain constraints. If we know the laws of these complex systems, other specimens on possibly other material basis become imaginable. AI research is interested in the theory of cognitive information systems to simulate specimens of biological evolution or to build new systems for technical purposes. During biological evolution on Earth, cognitive abilities were most differentially developed by the human brain. A distinction must be made between neuronal subsystems and areas that realize cognitive functions. In the vertebrate brain, five parts had developed approximately at the same time in phylogeny, namely myelencephalon, behind brain, midbrain, diencephalon, and cerebrum. The brain and spinal cord together form the central nervous system. The brain stem comprises the extended

© Springer-Verlag GmbH Germany, part of Springer Nature 2020
K. Mainzer, *Artificial intelligence – When do machines take over?*, Technik im Fokus,
https://doi.org/10.1007/978-3-662-59717-0_7

spinal cord (myelencephalon), the bridge (pons), the cerebellum and midbrain [2].

Background Information

Primates like humans are distinguished by the following parts and functions: The bridge (pons) transmits motion signals from the cerebral cortex to the cerebellum. The cerebellum regulates movements and is involved in learning motor skills. The midbrain controls sensory and motor functions. The diencephalon comprises the thalamus as the control center for signals from the rest of the central nervous system to the cerebral cortex and the hypothalamus as the regulator of vegetative, endocrine (i.e. glandular secretions) and visceral (i.e. intestinal) functions.

Brain stem, limbic system, and neocortex are not separate, but closely connected. The limbic system proves to be a processing and integration organ. It connects the frontal brain with deep brain structures, which are responsible for the control of vital functions such as blood pressure and respiration. We will see later that cognitive processes such as thinking and speaking cannot be separated from motivating emotions. In particular, motivating action, behavior, and target assessments are vital and excluded without feedback with the limbic system. The brain is therefore not a computer in which the neocortex could be separated as an arithmetic unit.

The cerebral cortex consists of a 2–3 mm thick nerve layer that covers the cerebral hemisphere. Each of the two halves of the cerebrum is divided into the four large areas of forehead or frontal lobe, vertex or parietal lobe, occipital lobe and temples or temporal lobe. The frontal lobe is primarily involved in the planning of future actions and movement coordination. The parietal lobe supports the sense of touch and the body perception, the occipital lobe the sight and the temporal lobe the hearing and partly the learning, the memory, and the emotion. The names of these regions are derived from the skull regions covering them.

In addition to the cerebral cortex, the cerebral hemisphere also includes the lower basal ganglia, the hippocampus and the amygdala. The basal ganglia participate in the motor control. The hippocampus is an ancient developmental structure in both temporal lobes of the cerebral cortex. The Latin name derives from the formation reminiscent of a seahorse. It plays a major role in learning processes and memory formation. The almond kernel formation coordinates vegetative and endocrine reactions associated with emotional states.

Perception of the outside world is a central capability of a cognitive information system. Human sensory organs register physical and chemical systems of the outside world, which are then processed by complex neuronal systems into what we call perceptions. Different forms of energy such as light, energy, mechanical, thermal or chemical energy are transformed into the five sensory qualities or modalities of seeing, hearing, feeling, tasting and smelling.

Background Information

The neuronal organization of the different perceptual systems is very similar:

First, primary sensory neurons are activated by stimuli. We also speak of receptor neurons, which are characterized by local receptive fields. In the sense of touch, each primary sensory neuron corresponds to a delimited receptive field on the skin, in which the respective neuron can be activated by pressure, for example.

The primary sensory neurons are bundled by second-order neurons in the CNS ("projection neurons"), which are further interconnected with superordinate neurons. Relay nuclei play a decisive role here.

As already mentioned, many relay nuclei are concentrated in the thalamus in order to transmit sensory signals to the cerebral cortex via specific nerve tracts. The receptive fields of the individual primary sensory neurons overlap to fields of the individual projection neurons, so that finally receptive fields of higher order neurons are also found in the sensory regions of the cortex.

The paths of the sensory systems are hierarchically organized from the receptors to the neurons of higher order with corresponding receptive fields. Partial modalities of the sensory systems such as, e.g., shape, colour, and movement in the visual system or touch, pain, and temperature in the sensory system of the body have separate and parallel paths, which are bundled on the hierarchy levels.

Finally, they converge in the respective sensory cortex areas in order to produce uniform sensations such as a red sweet apple or a burning pain. These perception systems therefore work with separate and parallel signal processing. The stimuli of neighboring receptors (e.g., on the skin or retina) are translated into neighboring signals of the receptive fields of higher order neurons.

The receptive fields are thus neuronal (topographic) maps in which the spatial order of the input signals is preserved at each hierarchical level of the sensory system. Only with the chemical senses such as taste and smell, this principle of order does not apply.

Just like perceptions, movements and emotions, cognitive processes such as memory, learning, and language are controlled by complex neuronal circuits in the brain. Learning is a process by which information systems acquire information about themselves and their environment. Memory is the ability to store and retrieve this information. Depending on the length of storage, we differentiate in humans between short-term memory, which comprises seconds to minutes, and long-term memory, which comprises days to decades.

In learning, we speak of an explicit form when data and knowledge are consciously acquired and kept permanently available. The implicit form is about the acquisition of motor and sensory abilities, which are available at all times without consciousness. For example, while driving a car, the driver acquires explicit factual knowledge in theoretical lessons, while driving practice begins with explicit instructions from the driving instructor, but it is ultimately essentially based on unconscious motor and sensory learning programs. Analogously, computer science distinguishes between declarative (explicit) and non-declarative (implicit) knowledge.

▶ Definition

Cognition research distinguishes between two declarative and two non-declarative memory systems [3]:

Episodic memory is responsible for autobiographical and individual events. It is just as declarative as the knowledge system, which explicitly stores facts and knowledge from textbooks.

In contrast, procedural memory encompasses motor abilities that do not require conscious representation of knowledge for their execution.

Often priming is also mentioned as an implicit form of a memory system, since it spontaneously and unconsciously associates similar experienced situations and patterns of perception with each other (In advertising, this form of memory is used to animate customers below the threshold of awareness for actions and decisions).

Cognitive memory systems require procedures for storing, storing, and retrieving information. In organic systems such as the human brain, neuronal subsystems can be differentiated that are responsible for this. Thus, the association regions of the cerebral cortex are used for storage in episodic memory and in the knowledge system. The cerebellum is involved in the storage of procedural memory, while in priming this task is performed by areas around the primary sensory fields of the cerebral cortex.

It is already remarkable at this point that emotion-processing areas of the limbic system can also be involved in the storage of information. For animal and human memory systems, it follows that emotions must often be addressed in order to facilitate memorization and learning. This aspect has hardly been considered for artificial memories and memory systems in AI research so far, but could play a role in the development of novel storage systems that react sensitively to the human user.

▶ **Definition**

In implicit learning, cognitive research distinguishes between associative and non-associative forms. A well-known example of associative learning is classical conditioning (z. B. Pavlovian dog), in which a temporal relationship (association) between a conditional stimulus (e.g.. sound signal) and a subsequent unconditional stimulus (e.g., food supply) is learned.

During operative conditioning (e.g., trial and error learning), a behavior (e.g., accidental finding and pressing of a button) is intensified by a stimulus (e.g., food).

In non-associative learning, repeated stimulus signals unconsciously produce an acclimatization with a decrease in reaction (habituation) or an increase in stimulation with an overreaction (sensitization).

Associative implicit learning such as classical conditioning can be explained by neuronal information processing. The basis is the synaptic amplification between sensory neurons, which are activated one after the other by a conditional and unconditional stimulus. These sensory neurons are interconnected via inter- or projection neurons. Synaptic enhancement is achieved when the interneurons are activated by the unconditional stimulus shortly after the sensory neurons stimulated by the conditional stimulus have begun to fire.

In fact, a conditioning training in coupled sensory neurons shows a greater excitatory (postsynaptic) potential than in uncoupled neurons. The explicit form of learning and memory is associated with long-term potentiation in the hippocampus. This molecular-biologically confirms a psychological rule of learning that was introduced as a hypothesis in 1949 by D. Hebb [4]:

▶ If an axon of cell A excites cell B and repeatedly and permanently contributes to the excitation of action potentials in cell B, the efficiency of cell A for the generation of action potentials in B increases (Hebb's rule).

7.2   Neural Networks and Learning Algorithms

W. S. McCulloch and W. In 1943 Pitts proposed a first model of a technical neural network [5]:

▶ Definition

In a simplified McCulloch-Pitts neuron, the dendrites are replaced by input lines $x_1 \ldots x_m$ ($m \geq 1$) and the axon by an output line y (Fig. 7.1).

If the input line x_i conducts a pulse in the nth time interval, $x_i(n) = 1$, otherwise $x_i(n) = 0$.

If the ith synapse is excited (excitatory), it is connected to a weight w_i greater than zero, which corresponds to the electrical strength of an electrical synapse or the transmitter output of a chemical synapse.

In the case of an inhibitory synapse, $w_i < 0$ applies.

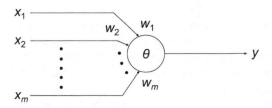

Fig. 7.1 McCulloch-Pitts-Neuron

If the time interval until the next impulse (refractory time) is interpreted as a time unit, a digital time scale $n = 1, 2, 3, \ldots$ can be assumed in which the neuron operates.

The firing of the output at time $n + 1$ is determined by the firing of the input signals at time n: The neuron fires according to McCulloch-Pitts a pulse along its axon at time $n + 1$, if the weighted sum of the input at time n exceeds the threshold value of the neuron.

A neural network according to McCulloch Pitts is understood as a complex system of such neurons whose input and output lines are connected with each other and which operate with the same time scale.

A major limitation of the McCulloch-Pitts nets was the assumption that the weights were fixed forever. Thus, a decisive performance capability of the brain is excluded from its evolution in phylogeny. Learning is made possible by modifications of the synapses between the neurons. Therefore, it requires variable synapse weights. The strength of the connections (associations) of neurons depends on the respective synapses. From a physiological point of view, learning is therefore a local process. The changes of the synapses are not caused and controlled globally from the outside, but happen locally at the individual synapses by changing the neurotransmitters.

According to this concept, the American psychologist F. Rosenblatt built the first neuronal network machine at the end of the 1950s, which was supposed to accomplish pattern recognition with neuron-like units:

Example

This machine, which Rosenblatt gave the name "Perceptron", consisted of a grid of 400 photocells modelled on the retina and connected to neuron-like units [6].

When a pattern such as a letter was presented to the sensors, this perception activated a group of neurons, which in turn caused a group of neurons to classify whether or not the letter presented belongs to a particular letter category.

Analogous to neuronal tissue, Rosenblatt provided for different layers:

The input layer serves as an artificial retina. It is composed of stimulus cells (S-units), which can technically be imagined as photocells.

The S-units are linked to the middle layer via random connections that have fixed weights (synapses) and are therefore not changeable or adaptive. According to its task Rosenblatt calls them association cells (A-units). Each A-cell thus receives a firmly weighted input from some S-cells. An S-cell of the retina can also project its signal onto several cells of the middle layer. The middle layer is completely connected to the response cells (R-units) of the output layer (Fig. 7.2).

Only the synapse weights between means and output layer are variable and thus capable of learning.

The neurons work as switching elements with two states. The perceptron network learns through a monitored learning process:

For each pattern to be learned (e.g. letter) the desired state of each cell of the output layer must be known. The patterns to be learned are offered to the network. The perceptron learning rule is a variant of Hebb's learning rule according

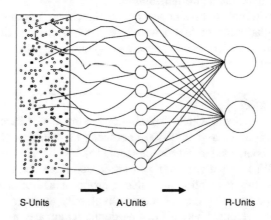

S-Units   A-Units   R-Units

Fig. 7.2 Architecture of Perceptron

to which the change in weight in one learning step is proportional to presynaptic activity and to the difference between desired and actual postsynaptic activity. The process is repeated with a constant learning step until all patterns produce the correct output.

However, due to its slowness and limited learning ability, in which only the synaptic weights can be changed to the output layer, the perceptron proved to be practically unusable. In addition, there was a serious mathematical limitation.

The perceptron learning algorithm (1950) begins with a random set of weights and modifies these weights according to an error function to minimize the difference between current output of a neuron and desired output of a trained data pattern (e.g., letter sequences, pixel image). This learning algorithm can only be trained to recognize supervised learning patterns that are "linearly separable". In this case, the patterns must be clearly separable by a straight line.

Example

Figure 7.3a shows two patterns consisting of either small squares or small circles as elements. Both patterns can be separated by a straight line and are thus recognizable by a perceptron. Figure 7.2b shows two patterns that cannot be separated by a straight line.

► M. Minsky, leading AI-researcher and S. Papert proved in 1969 that the perceptron would fail if the patterns were only separated by curves ("nonlinear") (Fig. 7.3b; [7, 8]).

Initially, Minsky and Papert regarded the proof as the fundamental limit of neural networks for AI research. The solution to the problem was inspired by the architecture of natural brains. Why should information processing only run in one direction through the superimposed layers of networked neurons? D. E. Rumelhart, G. E. Hinton, and R. J. Williams proved in 1986 that feedback

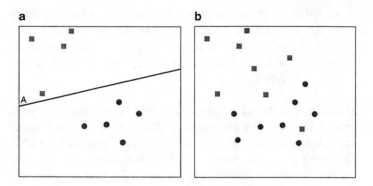

Fig. 7.3 Linear (**a**) and nonlinear (**b**) separable patterns

information flows (backpropagation) between input, intermediate, and output layer with suitable activation and learning algorithms also allow nonlinear classifications. In 1989, K. Hornik, M. Stinchcome, and H. White proved that under appropriate conditions also feedforward-architectures can be used [9, 10]:

▶ **Definition**
A nonlinear function of input variables is determined by optimizing the weights of the function y(W,X) with the weight vector W to be calculated, the vector X of the known inputs, and the known output y.

A (feedforward) neural network with 3 layers of input neurons, middle ("hidden") neurons and output neurons (Fig. 7.4) is determined by the output function

$$y(Z, W, X) = o(Z \cdot h(W \cdot X))$$

with input vectors X, weight vectors W between input layer and hidden neurons, activation function h of hidden neurons, weighting vector Z between hidden neurons and output neurons and activation function of the output neuron. With an output neuron, individual numerical values can be predicted.

Fig. 7.4 Three-layer model of a neural network with an output neuron

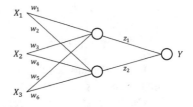

▶ **Definition**

A feedforward neural network with 3 layers and two output neurons is determined by the output

$$y_1(Z_1, W, X) = o(Z_1 \cdot h(W \cdot X))$$

$$y_2(Z_2, WX) = o(Z_2 \cdot h(W \cdot X))$$

function with weight vectors and between the hidden neurons and the two output neurons (Fig. 7.5).

Multiple output neurons can be used for classification tasks. Neural networks learn to predict to which class (corresponding to the number of output neurons) an input belongs (e.g. face, profile recognition) [11].

Multilayer neuron networks are used in visual perception. They can be simulated on the computer:

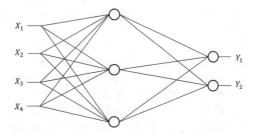

Fig. 7.5 Three-layer model of a neural network with two output neurons

Example

On the first layer, the computer identifies brighter and darker pixels.

On the second layer, the computer learns to identify simple shapes such as corners and edges.

On the third layer, the computer learns to distinguish more complex parts such as details from faces.

On the fourth shift, the computer learns to assemble the parts into faces.

In fact, the steps of facial recognition are performed in this order from the retina to the visual fields of the human brain (Fig. 7.6; [12]).

Learning processes with multi-layered neural networks are also referred to as "deep learning". What is meant here is that a step-by-step "deeper" understanding of a factual situation (e.g., a picture) develops after first only individual building blocks, then clusters and finally the whole are recognized.

At the beginning of the 80s, the physicist J. Hopfield developed a single-layer neural net, which can be applied to the model of self-organizing materials (spinglas model) [13].

▶ **Definition**

A ferromagnet is a complex system of dipoles, each with two possible spin states up (↑) and down (↓). The statistical distribution of the up and down states can be specified as the macro state of the system.

When the system cools down to the Curie point, a phase transition takes place in which almost all dipoles spontaneously jump into the same state and therefore a regular pattern emerges from an irregular distribution. The system thus changes into a state of equilibrium, in which an order is organized independently. This order is perceived as the overall magnetic state of the ferromagnet.

Hopfield analogously described a network of a single layer in which binary neurons are completely and symmetrically

Fig. 7.6 Multilayer model for face recognition (deep learning) [12]

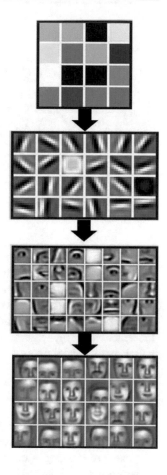

networked. It is therefore a homogeneous neural network. The binary state of a neuron corresponds to the two possible spin values of a dipole. The dynamics of the Hopfield system is modelled exactly on the spin glass model of solid state physics. The energetic interaction of the magnetic atoms in the spin glass model is now interpreted as an interaction of binary neurons. The distribution of the energy values in the spin glass model is understood as the distribution of the 'computing energy' in the neuronal network.

Fig. 7.7 Potential mountains as state space of a Hopfield system

We can clearly imagine a potential mountain range above the state space of all possible binary neurons (Fig. 7.7) [14]. If the system starts from an initial state, it moves downhill in this potential mountain range until it gets stuck in a valley with a local minimum. If the starting state is the input pattern, then the reached energy minimum is the response of the network. A valley with a local energy minimum is therefore an attractor to which the system moves.

Example

A simple application is the recognition of a noisy pattern whose prototype the system has learned before. For this purpose, we imagine a chessboard-like grid of binary technical neurons (Fig. 7.8) [15]. A pattern (e.g., the letter A) is represented in the grid by black points for all active neurons (with value 1) and white points for inactive neurons (with value 0). The prototypes of the letters are first "trained" to the system, i.e. they are connected to the point tractors or local energy minima. The neurons are connected to sensors that perceive a pattern.

If we now offer the system a noisy and partially disturbed pattern of the trained prototype, it can recognize the prototype in a learning process:

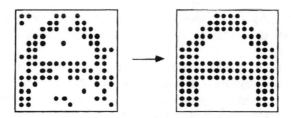

Fig. 7.8   Recognition of a pattern in the Hopfield system

The learning process takes place through local interactions of the individual neurons according to a Hebb's learning rule. If two neurons are either active or inactive at the same time, the synaptic coupling is amplified. In different states, the synaptic weights are reduced. The learning process continues until the stored prototype is generated ("recognized").

The recognition process is therefore a phase transition to a target attractor, as we have already observed with ferromagnets. It should be emphasized that this phase transition happens without central program control by self-organization. Hopfield systems can also be used for cognitive tasks. In the potential mountains, the lowest energy state represents an optimal solution. An optimization problem, such as the search for an optimal travel route at is also possible with a Hopfield system:

Example

In a Hopfield system, the different distances between cities and the order in which they are visited are taken into account by corresponding synaptic weights.

In fractions of a second, the computing energy sinks into a stable, low-energy state that represents the shortest route.

A neural network can thus constantly decide between millions of possible answers, because it does not have to check the answers one after the other. Nor does it assume that every possible answer is true or false. Each possibility rather has its synaptic weight, which corresponds to the strength of the assumption that connects the system with each possibility. They are processed in parallel.

Hopfield systems work in parallel, but deterministically, i.e. every neuron is indispensable for character recognition. However, living nerve cells hardly behave like determined planetary systems, and even with corresponding technical network models, major disadvantages occur. If we imagine a recognition process or a decision-making process in the sense of Hopfield as energy reduction, then the learning process can get stuck in a valley that is not the deepest in the whole network.

T. J. Sejnowski and G. E. Hinton therefore proposed a procedure to lead the network to ever deeper valleys [16]. If a sphere has reached a valley in the energy mountains, then the obvious and descriptive suggestion is to shake the whole system a little such that the sphere can leave the valley to take lower minima. Strong or weaker shaking movements change the probability of a sphere being located as with a gas molecule, whose collisions are influenced by pressure and temperature changes.

▶ Therefore, Sejnowski and Hinton named their probabilistic network "Boltzmann machine" after the founder of statistical mechanics and thermodynamics.

It is remarkable that John von Neumann had already pointed out the connection between learning and cognitive processes to Boltzmann's statistical thermodynamics.

The problem of finding a global minimum in the network and avoiding secondary minima occurs physically in the thermodynamics of crystal growth. In order to give a crystal a structure that is as free of defects as possible, it must be cooled slowly. The atoms must have time to find places in the lattice structure with minimal total energy. At sufficiently high temperatures, individual molecules are able to change their state in such a way that the total energy increases. In this case, local minima can still be left. As the temperature drops, however, the probability of this happening decreases. This procedure is also clearly described as "simulated annealing or cooling".

Probabilistic networks are experimentally very similar to biological neuronal networks. If cells are removed or individual synaptic weights are changed by small amounts, Boltzmann machines prove to be the right choice as fault-tolerant towards smaller disturbances such as the human brain with smaller accident damages. The human brain works with layers of parallel signal processing. For example, internal intermediate steps of neuronal signal processing that are not in touch with the outside world are taken between a sensory input layer and a motor output layer. In fact, the representation and problem-solving capacity of technical neural networks can also be increased by interposing different layers with as many neurons as possible that are capable of learning. The first layer receives the input pattern. Each neuron of this layer has connections to each neuron of the next layer. The series of executive transactions one after the other continues until the last layer is reached and gives off an activity pattern.

▶ Supervised learning procedures mean that the prototype to be learned (e.g., the recognition of a pattern) is known and the respective error deviations can be measured. A learning algorithm must change the synaptic weights until an activity pattern emerges in the output layer that deviates as little as possible from the prototype.

▶ An effective method is to calculate the error deviation of actual and desired output for each neuron in the output layer and then trace it back through the layers of the network. That procedure is called a backpropagation algorithm. The intention is to reduce the error to zero or negligibly small values by a sufficient number of learning steps for a default pattern.

So far such a method has been assumed to be technically effective, but biologically unrealistic, as neuronal signal processing is only forward (feedforward) from presynaptic to postsynaptic neuron. In the case of long-term potentiation, however, reverse signal effects are also discussed today. Therefore, learning algorithms with backpropagation could become neurobiologically interesting.

▶ Unsupervised learning means that a learning algorithm recognizes new patterns and correlations without resorting to predefined or trained prototypes

How can a neural network learn without "being supervised" by an external instance (prototype or "teacher")? Highly developed brains of biological evolution can not only recognize trained patterns, but spontaneously classify them according to characteristics without external monitoring of the learning process. Terms and figures are generated by spontaneous self-organization. In a multi-layered network hierarchy, this process can be realized by competition and selection of neurons in the different layers. A neuron learns by winning the competition with the other neurons of a cluster. Similarities between correlations and contexts are reinforced (Fig. 7.9; [17]).

In fact, sensory information is projected across multiple neuronal layers in the cortex. Although these projections are distorted, they preserve the neighborhood relationships between points of the depicted object. Such distorted representations can be found in the visual fields of visual perceptions as well as in the somatotopic representations of the body surface or the auditory fields of tone sequences in the cortex.

The visual, tactile, or auditory images of the outside world may be cut into pieces to better fit on the rutted surface of the cortex. They may also be larger or smaller distorted to express greater or lesser sensitivity in certain zones through greater or lesser resolution. In these parts, however, the order of the correlations is preserved. Therefore, perceptual information is not only projected over several neural layers and processed in parallel. In the neuronal layers, neighbouring neurons of a neuronal layer are also influenced.

Neuronal projections in the cortex are called images and neuronal maps because of their geometric properties. They organize their detailed representations largely by themselves due to external stimuli. This neuronal self-organization relieves genetic information processing. Not all details can be considered because of their permanent change and complexity in the programming of genetic information. For visual maps, the

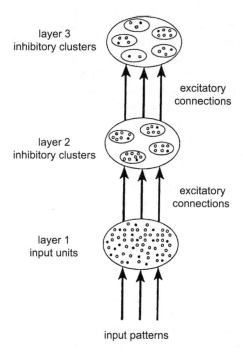

layer 3
inhibitory clusters

excitatory
connections

layer 2
inhibitory clusters

excitatory
connections

layer 1
input units

input patterns

Fig. 7.9 Multi-layer neural network detects correlations and clusters

computational neuroscientist C. von der Malsburg proposed an
algorithm of self-organization [18]:

▶ **Definition**
In neuronal maps, synapses change their connection due to the
activity of their connecting partners (Hebb's rule).

Neuronal signals are not only transferred to subsequent neuronal layers, but also within a layer to neighbouring neurons.

Neurons compete with each other, with the more strongly
activated neurons suppressing the weaker ones.

Since many neuronal maps work in this way, it is assumed that
only the principles of self-organization are genetically anchored,
but that the details develop independently in each individual.

Following this biological model, T. Kohonen has designed neuronal networks as self-organizing feature maps that make do with unsupervised learning algorithms [19]. Neighbouring excitation sites of the characteristic map correspond to external stimuli with similar characteristics. These include stimulus sites on the retina, skin, or in the ear that are mapped on a neuronal layer in the cortex.

▶ **Definition**
Schematically, the input signals of the stimuli (geometrically interpreted as vectors of a vector space) are mapped onto a quadratic grid in a Kohonen map, whose nodes represent the neurons of the cortex (Fig. 7.10).

In a single learning step, a stimulus is randomly selected and mapped to the best adapted neuron. This is the neuron whose synaptic strength differs least from the input stimulus compared to the synaptic strengths of the other lattice neurons.

All neurons in the vicinity of this excitation center are also excited, but less with decreasing distance. They adapt to the excitation center in the learning step.

However, the learning algorithm depends on both the range of the neighbourhood relationship and the reaction intensity to new stimuli. Both values decrease with each repeated learning step until the map has changed from a disordered initial state to a feature map whose details correspond as adequately as possible to the distribution of the input stimuli. The information system has

Fig. 7.10 Kohonen map
with unsupervised learning
[19]

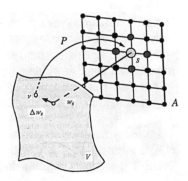

independently entered a state of equilibrium, which is connected with the generation of information.

In unsupervised learning, the algorithm learns to recognize new patterns (correlations) from the set of inputs without "teachers" (e.g., self-organizing Kohonen maps). In supervised learning, the algorithm learns to determine a function from given pairs of inputs and outputs (training). A "teacher" (e.g., trained prototype of a pattern) corrects deviations from the correct function value to an output (e.g., recognition of learned patterns).

Reinforcing learning is in between: A robot is given a target (as in supervised learning). However, it must find the realization independently (as in unsupervised learning). In the step-by-step realization of the goal, the robot receives feedback from the environment at each partial step as to how good or bad it is at realizing the goal. Its strategy is to optimize this feedback.

Technically this means: The algorithm learns through experience (trail and error) how to act in an (unknown) environment (world) in order to maximize the utility of the agent [20, 21].

▶ **Definition**
Mathematically, reinforcement learning is a dynamic system of an agent and its environment with discrete time steps $t = 0$, 1, 2, ... At any time t the world is in a state z_t. The agent chooses an action a_t. Then the system enters the state z_{t+1} and the agent receives the reward b_t (Fig. 7.11).

The agent's strategy is defined with π_t wherein $\pi_t(z, a)$ is the probability that the action is $a_t = a$ if the state is $z_t = z$.

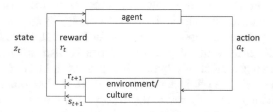

Fig. 7.11 Reinforcing learning of an agent from its environment

Algorithms of reinforcing learning determine how an agent changes its strategy based on its experiences (rewards). The goal of the agent is to optimize his feedback in order to achieve the goal.

Example

An example is a mobile robot who's supposed to pick up empty beverage cans in an office and throw them in a trash can. The robot has sensors to detect the cans and an arm with a gripper to grip the cans. Its activities depend on a battery that occasionally needs to be recharged on a base station. The control system of the robot consists of components for the interpretation of sensor information and for the navigation of the robot arm and robot gripper. The intelligent decisions for can search are realized by a reinforcement algorithm that takes into account the charge level of the battery.

The robot can choose between three actions:

1. active search for a can in a certain time period,
2. stationary hibernation and wait for someone to bring a can,
3. back to base station to recharge the battery.

A decision is made either periodically or whenever certain events such as finding a can occur. The condition of the robot is determined by the condition of its battery.

The rewards are usually zero, but become positive when the robot finds an empty can, or negative when the battery charge runs out.

Ideally, an agent is in a state that sums up all the past experiences necessary to achieve its goal. Normally its immediate and present perceptions are not sufficient for this. But, more than the complete history of all past perceptions is also not necessary. For the future flight of a ball, it is sufficient to know its current position and speed. It is not necessary to know the complete previous course. In such cases, the history of the present state has no influence on future development. If the probability of a state depends only on the preceding state and a preceding action of the agent in that state, the decision process satisfies the Markov property:

▶ **Definition**

The Markov Decision Process (MDP) is determined by the Markov property:

$$P(z_{t+1}, r_{t+1} | z_{0:t}, a_{0:t}, b_{0:t}) = P(z_{t+1}, r_{t+1} | z_t, a_t)$$

The action model $P(z_{t+1} | z_t, a_t)$ is the conditional probability distribution that the world changes from state z_t to state z_{t+1} if the agent selects the action a_t; r_{t+1} is the expected return in the next step.

Because computing and storage capacity are scarce and costly, practical reinforcement learning applications often require the Markov feature. Even if the knowledge of the present state is not sufficient, an approximation of the Markov property is favorable. For very large ("infinite") state spaces, the utility function of an agent must be approximated (e.g., SARSA = State-Action-Reward-State Algorithms, temporal difference learning, Monte Carlo methods, dynamic Programming).

According to the English mathematician and theologian T. Bayes (1702–1762), learning can be explained by conditional probabilities of events. Probability is not understood as frequency (objective probability), but as degree of belief (subjective probability): An event A is set before the occurrence of event B with the a priori probability $P(A)$, but after the occurrence of B with the a posteriori (conditional) probability $P(A|B)$.

Background Information

The Bayes theorem can be used to calculate conditional probabilities: The conditional probability $P(A|B)$ of event A after the occurrence of an event B is determined by the quotient of the probability $P(A \cap B)$ (i. e. the probability that the events A and B will occur together) and the probability of $P(B)$ of event B

$$P(A|B) = \frac{P(A \cap B)}{P(B)}$$

$$= \frac{\frac{P(A \cap B)}{P(A)} \cdot P(A)}{P(B)} = \frac{P(B|A) \cdot P(A)}{P(B)}$$

Therefore, the theorem of Bayes says $P(A|B) = \frac{P(B|A) \cdot P(A)}{P(B)}$

i. e. the probability of A after the occurrence of B is calculated from the conditional probability of B provided that A and the a priori probabilities $P(A)$ and $P(B)$ are given.

Learning from experience can be realized by learning theorems with conditional probabilities. An artificial intelligence could use this procedure to assess future decisions and actions:

▶ A Bayesian net consists of nodes for event variables whose connections (edges) are weighted by conditional probabilities. From this the probabilities of events under the condition of other events can be calculated.

Example

Events E (earthquake) and B (burglary) trigger event A (alarm sounds) (Fig. 7.12; [22]). The alarm causes John (event J) or Mary (event M) to call the fire brigade.

The variables E (earthquake), B (burglary), A (alarm sounds), J (John's calls), M (Mary calls) are binary and linked to the truth values T (event occurs) or F (event does not occur). Therefore, alarm A can be caused by an earthquake E or a burglary B. The alarm can trigger a call from John J or Mary M to the fire brigade.

If for example, a break-in is observed and there is no earthquake, John calls and not Mary: How high is then the probability that alarm sounds?

Bayesian networks allow effective forecasting and decision models. However, human decision-making processes are rarely "rational". They are often "distorted" by feelings and intuitions. Psychologist and winner of the Nobel Prize for Economics D. Kahneman draws attention to this and speaks of cognitive distortion of human choices [23]. Often, these allegedly "typically

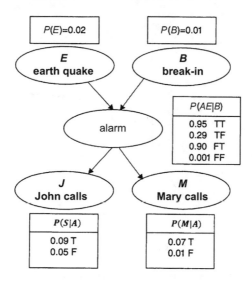

Fig. 7.12 Bayesian learning net with conditional probabilities [22]

human" reactions were cited in order to reduce artificial intelligence against human intelligence after the Turing test.

A rationally decisive artificial intelligence can be described according to the classical utility theory.

In classical utility theory, an expectation value is calculated by multiplied the possible results of $x_1, x_2, ..., x_n$ with their likelihood of occurrence $p_1, p_2, ..., p_n$ and then added up the weighted results $p_1x_1, p_2x_2, ..., p_nx_n$.

Formally, "cognitive distortions" of rational expectation values can be considered in Artificial Intelligence:

To calculate the expected value u of a utility, Kahneman considers the cognitive distortions of probabilities of occurrence and results through a value function v in $u = \sum_{i=1}^{n} w(p_i)v(x_i)$. The function chosen for this (see Fig. 7.13) is non-linear S-shaped and weights losses more than profits. A weighting function w of probabilities takes into account that people overestimate unlikely outcomes and underestimate more likely outcomes.

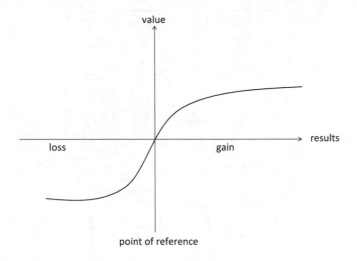

Fig. 7.13 Cognitive distortion of rational decisions

One example is fear of flying, where a rare event such as a plane crash is overrated. In contrast, much more frequent road accidents are underestimated. Therefore, a robot could be programmed with a function of cognitive distortion to pass the Turing test in this case. But, it doesn't "have" any emotions yet, but only simulates them.

7.3   Emotions and Consciousness

Just like perceptions and movements, emotions are controlled by neuronal circuits in the brain. Emotions are therefore also based on signals and information processes that can be made accessible to artificial intelligence.

In biological organisms, neuronal messengers play a role in the emotional circuits. Already with animals it is to be observed that fear is connected with external reactions like, e.g., acceleration of heartbeat and respiration or dry mouth. The emotional system is interconnected with the responsible autonomous nervous system. But, cognition and memory also have an effect on emotional states: Joy, sadness, or pain can be associated with ideas and memories. Brain research shows how closely thinking, feeling, and acting are networked in humans. Psychology, therefore, also speaks of an emotional intelligence of the human being, which is typical for his decisions. The aim of AI research is to technically model or even generate information systems with emotional intelligence. In this case, both cognition and emotion would not be limited to biological organisms.

Background Information

For a modelling, the external reactions in the case of emotional processes. Physiological changes in anxiety states such as rapid heartbeat, skin transpiration, or tense muscles are triggered by the autonomous nervous system, which is not consciously experienced, but reacts involuntarily (autonomously).

In the diencephalon, the hypothalamus serves as the control center. It registers changes in external and internal states and uses the autonomous nervous system to adjust and stabilize the body to new situations. Increasing the heartbeat to increase blood supply or dilating the pupil for rapid reactions are examples of alarming conditions such as anxiety. In addition, the hypothalamus acts on the (endocrine) gland system to release hormones. Therefore, the hypothalamus can be compared with an equilibrium regulator in a homeostatic system. By electrically simulating the hypothalamus in cats and rats, states can be generated that are associated with typical external reactions such as anger.

Where do emotions arise in the human brain? A ring of cerebral cortex around the brain stem and the diencephalon (limbic lobe) is involved. The limbic system also includes hippocampus and amygdala. In fact, the failure of the temporal lobe with a tonsil and hippocampus formation led to "emotional blindness": patients show no emotional expressions.

In addition to compounds of the amygdala, interconnections with the association cortex are indicated. In detail, the neuronal switching units of the amygdala structure are networked in a complex and large-scale manner with various neuronal systems, which to date are only partially known. So, there are impulses from the sensory nuclei of the thalamus and from the primary sensory cortex. Input and output are fed back to the cortical

association fields through a switch core of the amygdala structure (nucleus centralis) and thus enable the conscious experience
of emotions. This switch core is also involved in the degree of
alertness and the physiological reactions associated with it.

The amygdala structure is also involved in the emotional staining of cog-
nitive-sensory signals. The limbic system crosses certain transmitter path-
ways that are involved in generating aggression, anxiety, sad, and depressive
feelings. Morphines generated by the brain cause joyful and lustful feelings.
Emotional states are thus caused by different, widely interconnected brain
structures. Known are the neural anger-fury, fear-anxiety, panic-sadness,
joy-lust and interest-expectation systems.

The neurologist A. Damásio distinguishes between a basic apparatus of primary feelings due to congenital neuronal circuits of
the limbic system and secondary feelings acquired in individual
development due to special experiences [24]. Secondary feelings
arise from the modification and further development of primary
feelings by connecting their basal neural circuits with the pre-
frontal cerebral cortex, thus enabling individual experiences,
memories, and learning processes. As with sensory and motor
systems, the emotional system does not have a fixed emotional
base map, but a multitude of neuronal representation patterns
that are permanently modified and coordinated.

▶ **Definition**
This dynamic of emotional states is the research topic of affective computing, which draws on the methods of artificial intelligence [25]. The first priority is to improve the human-machine
relationship. An example is training neural networks to recognize emotional reactions.

The improvement of the interface between computer and user
should lead to a computer without mouse and keyboard of a keyboard being operated by facial expressions, gestures, or vocal
pitch. Disabled people in particular could benefit from this.

It is assumed from brain research that emotions are represented
by physiological signal patterns that can be recognized by a neuronal network. For example, anger or sorrow are determined by

certain measurement curves for muscle tension, blood pressure, skin conductivity and respiratory frequency. A neural net can be trained to recognize typical patterns in order to recognize basic moods even in noisy patterns. Such a neural network could be just as mistaken in its emotional diagnosis as a human psychologist who misinterprets these curves. It could therefore, in the spirit of Turing, take the Turing test, because erring is known to be human. On the other hand, the emotional recognition of a software would be realized by means of measuring instruments and quantities that are not decisive for the human perception of emotions.

Therefore, other approaches take into account the facial expression that is fundamental to the emotional body language of primates. Thus, there are a number of basic patterns of facial expressions, which are to be understood as codes of this language. There are several possibilities for a neural pattern recognition system.

Emotional facial expressions from, e. g., "happiness", "surprise", "anger", "disgust" are associated with tense or relaxed facial areas, which form heat or energy maps due to their greater or weaker blood circulation. Using special sensors, neural networks can perceive such emotional maps. The neural net is trained on corresponding prototypes and is thus able to recognize moods even in noisy patterns. In this case, the recognition system for emotion patterns works again differently than, for example, in humans. Instead of warming maps, which cannot be recognized by people without additional instruments, the facial expressions and skin colouring of typical parts of the face take place.

Example

Visual face maps can be recognized and distinguished by suitable neural networks. As an example, eight parameters for facial regions are distinguished: size, inclination, and shape of the eyebrows, size, width and shape of the mouth. Since these characteristics are more or less pronounced, they are defined as fuzzy properties in an interval from zero to one.

As far as the eye size is concerned, eyes can be closed (0), weak (0.33), moderate (0.67) or wide open (1.0). From the e-mail language, binary codes such as :-) or :-(are known for joy or frustration. In a fuzzy code, the mouth bow can vary between these two extreme positions as a representation of the shape of the mouth. An entire facial expression is thus determined by eight fuzzy values of these parameters. Six emotional parameters for happiness, sadness, anger, disgust, fear, and surprise can also be proven by fuzzy values between "weak", "moderate" and "strong". Then, a total emotional state is determined by eight fuzzy values of these parameters. As part of psychological tests, subjects can fill out test forms with fuzzy information about these parameters to determine their emotional state in certain situations. Situations are represented in this way by fuzzy values of an emotional pattern.

A Japanese research team around T. Onisawa has designed a system for situation-dependent facial recognition of emotions [26]:

Example

Emotional prototypes are based on happiness, sadness, anger, disgust, fear, surprise, and unnaturalness. The system consists of seven neural networks that determine the degree of each prototype.

Each net therefore consists of an input layer with eight nodes for facial parameters and six nodes for situation patterns and an output layer that receives one node for each of the seven emotional prototypes.

Between input and output layer there are two further neuron layers with twenty neurons each.

A backpropagation algorithm determines the deviation from the emotional prototypes. The same facial expression can be associated with different emotional states in different situations.

The seven output neurons of the overall system for the seven emotional prototypes describe the fuzzy value of an overall sensitivity.

If, for example three or more of the unpleasant emotions such as anger, disgust, sadness or fear are felt only a little, there is already a condition, which is described in natural language with the fuzzy term "malaise". Although this system is adaptive, flexible, fault-tolerant and fuzzy, it works on a different basis than the neuronal areas in the brain that are responsible for this task. There are no separate areas in the brain that can do nothing but recognize a particular type of emotion. On the other hand, the system takes into account cultural peculiarities that play a role in emotional sensitivities. Therefore, it is referred to by the Japanese word for feeling as a Kansei information system.

Can dynamic systems be designed that not only recognize emotions, but also feel them? Emotions in primates are associated with neurochemical and hormonal changes in complex networks. Emotions and sensations are not programmed in a cell or brain module as in the processor of a conventional computer, but in the synaptic circuit patterns and hormonal feedback of these networks. Circuit patterns that simulate emotional dynamics using learning algorithms are therefore conceivable.

Example

The CATHEXIS model assumes a network of emotional prototypes represented as nodes [27]. These are the seven emotions anger, fear, pain, sadness, pleasure, happiness, disgust, and surprise. Further emotions can be generated, again, as mixed states of the emotional network, in which the emotional prototypes are involved to varying degrees. For example, grief is a certain form of sadness, in which anger and fear resonate. The intensity of an emotional prototype is reinforced or inhibited by the intensity of the other prototypes. For example, anger inhibits pleasure and happiness, but intensifies sadness.

This emotional system is also equipped with a behavioral system in which different behavioral strategies can be selected. They range from facial expression and posture to vocal changes. Which behavior is chosen depends on the dominant emotions in each case. With corresponding numerical representations, an algorithm calculates the highest values that determine the behavior to be selected. This behavior is realized by a motor system in order to change the system environment.

The CATHEXIS model distinguishes four types of internal stimuli: Neuronal stimuli concern neurochemical and hormonal messengers. Senso-motoric stimuli concern muscle potentials and motor nerve stimuli. Motivation stimuli refer to all motivations that can trigger emotions. These include hunger and thirst as triggers of hunger and thirst as well as pain irritations as triggers of feelings of pain. Finally, cortical stimuli are assumed that can trigger emotions through thinking and decision making.

▶ Definition
The CATHEXIS system is algorithmicly defined through functions for calculating the intensity of emotional prototypes p at a point t. These functions depend on the intensity of an emotion at the previous time t − 1, the stimulating and inhibitory influence of the remaining emotional prototypes on p, and the overall impact of p by the four internal stimuli that simulate networking with the entire organism. Each emotion is determined by a lower threshold of triggering and an upper threshold of saturation.

This is how happiness and feelings of pain begin at a certain stimulus threshold and cannot be increased arbitrarily. These threshold values of individual emotions can, however, be different in humans. Some are more sensitive to pain, suffering, or euphoria than others.

It also makes sense to adjust the desired temperament of the emotional system by the respective threshold values of the emotional prototypes. Therefore, the dynamics of the emotional system CATHEXIS is determined by seven coupled equations for the seven assumed basic emotional types.

As a nonlinear complex dynamic system, it generates total states of the system, which stabilize in stationary equilibrium states, but also can crash into chaos. We can speak of emotional attractors that characterize the respective type of temperament—from the sanguine who oscillates between states such as "sky-high cheering" and "grieving to death," to the choleric who gets lost in a chaos attractor of anger.

Hybrid systems of artificial intelligence, which, in addition to motor and cognitive functions, can also be equipped with emotional subsystems. With respect to the brain, a knowledge-based expert system could be coupled with an emotional system analogous to the networking of the cortex with the control centers of the limbic system (e.g., the amygdala). The translation of this dynamic into a suitable programming language initially provides only an emotional software that does not feel itself. It could be sufficient for robots that corresponding actions are triggered at certain threshold values in numerical intervals of stimuli: Robots do not need to actually feel pain and pleasure to interact with emotional intelligence in a user-friendly way.

Thus, the experience of emotions with robots is by no means excluded, if the software of emotions would be connected with the corresponding wetware of hormonal, neurochemical, and physiological processes as with biological organisms. Even with this physiological-biochemical equipment, a robot would not have to feel like humans in every respect. In any case, the generation of a sentient system is not excluded in principle (cf. Chap. 8).

Most body and brain functions, perceptions and movements are unconscious, procedural, and non-declarative. In evolution, attention, alertness, and awareness have emerged as advantages of selection in order to act more cautiously and unerringly in critical situations. Therefore, awareness reduces uncertainty and thus contributes to the information gain of a system. However, a complex information system would be completely overtaxed if all its process steps were brought into "awareness" in this controlled way.

Even highly complex cognitive processes can occur unconsciously. Often we do not know how ideas and information for problem solutions came about. Technology, science, and cultural history are full of anecdotes from great engineers, scientists, musicians or literary figures who literally report on intuitive and unconscious ideas in their sleep. Even managers and politicians often make intuitive decisions without having consciously calculated through all the details of a complex situation. For AI research it follows from this that consciousness functions can play an important role for cognitive and motor systems, but by no means possess the constitutive function without which intelligent problem solutions would not be possible.

▶ In brain research, consciousness is understood as a scale of degrees of attention, self-perception and self-awareness. We first distinguish visual, auditory, tactile, or motor consciousness and by this we mean that we perceive ourselves in these physiological processes. Then, we know that we now see, hear, feel or move without always having to be aware of visual, auditory, tactile or motor processes.

The neurobiological explanation of conscious visual perception again relies on hierarchical models of parallel signal processing:

Example

At each hierarchical level, visual signals are coded anew and often differently on parallel paths. The ganglion cells of the retina process a light stimulus into action potentials. The neurons of the primary visual cortex respond differently to lines, edges, and colors.

Hierarchically higher neurons react to moving contours.

At even higher hierarchical levels, whole figures and familiar objects are coded, emotionally colored, and associated with memories and experiences.

Finally, premotoric and motoric structures are projected, whose neurons trigger activities such as speech and action.

This model explains why patients whose neural hierarchy level was destroyed for explicit shape perception no longer consciously recognize familiar faces, although they implicitly perceive a face with its typical details (contours, shadows, colors, etc.). Neurons that specialize in the perception of shapes (e.g., ompletion of contours, foreground-background) create the idea of figures, although these figures are only hinted at or suggested in an image.

Some philosophers taught about the "intentional" (intended by consciousness) relationship between the "cognitive subject" (observer) and the "cognitive object" (physical image). The development of the shape is missing in patients with a corresponding brain lesion. In injuries of a different hierarchical level, affected patients lose the ability to consciously perceive color, although the color receptors of the eye function.

The model of parallel signal processing at hierarchical levels of complex neural systems is of considerable importance for the technology of neural networks (see Sect. 7.2). For neurobiologists and brain researchers, however, it remains only a model as long as the neuronal structures involved and their molecular and cellular signal processing are not identified and proven by observation, measurement, and experiment.

In connection with states of consciousness, the actual problems of modern neurobiology, cognition, and brain research lie here. How are the neurons of a certain hierarchy level "interconnected" on a cellular level, which react to certain contours and shapes? Following Hebb's rules, simultaneous activity should not only excite the neurons that react to the respective aspect of a perceived object. Temporarily, the affected synapses would also have to be strengthened, so that a reproducible activity pattern develops in a kind of short-term memory.

In the systems of perception, we have already become acquainted with the synchronization process. According to this, all neurons that represent a certain aspect would have to fire in unison, but asynchronously to those that react to another aspect:

Questions

In continuation of this approach, the hypothesis could be developed that attention and states of consciousness would be generated by certain synchronous activity patterns (e.g., the attention to the foreground-background relationship in shape perception).

In this context, reference is made today to the aforementioned long-term potentiation of synaptic circuits, which is supposed to play a role in memory formation. It could guarantee the short or long-term reproducibility of conscious perceptions.

Other authors like F. Crick suggest that the neurons of a particular cortical layer are closely involved in states of consciousness by providing for the maintenance of circuits with circular excitation and attention.

Finally, there is the question of the emergence of a self-consciousness of ourselves, a self-consciousness that we call the "I". In the development of a child, the stages in which ego consciousness awakens and perceptions, movements, feelings, thoughts, and desires are gradually connected with one's own ego can be specified precisely. This does not switch on a single "consciousness neuron" like a lamp all at once. Such an idea would only postpone the problem, since we would have to ask how consciousness would come about in this neuron of consciousness. The example of perception illustrates the complex interconnection process that ultimately leads to self-confidence in the process of self-reflection: I perceive an object; finally, I perceive myself perceiving this object; finally, I perceive myself perceiving myself perceiving this object, etc. I perceive myself perceiving myself perceiving this object.

Questions

Each of these levels of self-perception could (according to a hypothesis of H. Flohr [28]) be connected with a certain neuronal representation pattern (neuronal map), the coding of which as input at the next level of perception triggers a new neuronal meta-representation of self-reflection.

Conversely, a stored activity pattern can be called by ourselves with the linguistic codeword "I" in order to express intentions and desires directly or to realize them through actions. This type of self-perception can be slowed, clouded, or accelerated to euphoric intoxication by medication and drugs. The synaptic interconnection speed in the formation of synchronous activity patterns can actually be influenced by the transmitter release.

The disposition to develop an ego-consciousness is probably genetic, even if we do not yet know exactly how. In evolution, it has evolved from paying attention to vital aspects of perception, movement, emotion, and cognition. We are already talking about historical, social and societal consciousness and by this we mean attention to important aspects of collective coexistence and survival. Also, for this kind of collective self-consciousness, complex networks of neuronal systems of perception, analysis, decision, but also of emotional evaluation and motivation for action are activated.

So if the laws that lead to complex brain states such as "consciousness" should be just as well known as the complex dynamics of another organ (e.g., heart), then complex systems with corresponding states could in principle not be ruled out. For these systems, the inner self-perception would also not necessarily be bound to linguistic representations.

Simple forms of self-monitoring are already implemented in existing computers and information systems. In biological evolution, animals and humans have developed forms of consciousness of growing complexity. If consciousness is nothing other than a special state of the brain, then "in principle" it cannot be seen why only the past biological evolution was able to produce such a system. The belief in the uniqueness of the brain's biochemistry is little supported by our technical experience to date.

After all, we humans were able to fly without feathers and flapping wings after the hydrodynamic laws of flying were known. Whether we speak of "artificial consciousness" in such complex systems, which would arise under suitable laboratory conditions, would then be just as much a question of definition as in the case of "artificial life". Within the framework of AI research, sensory, cognitive, or motor systems in a hybrid circuit could be connected to self-monitoring modules [29]. To what extent, however, such systems should be equipped with the capacities of attention, alertness, and awareness is not only a question of technical feasibility, but also, when the time comes, a question of ethics.

References

1. Mainzer K (1997) Gehirn, Computer, Komplexität. Springer, Berlin
2. Roth G (1994) Das Gehirn und seine Wirklichkeit. Kognitive Neuropsychologie und ihre philosophischen Konsequenzen. Suhrkamp, Frankfurt
3. Markowitsch HJ (1992) Neuropsychologie des Gedächtnisses. Verlag für Psychologie, Göttingen
4. Hebb DO (1949) The organisation of behavior. A neurophysiological theory. Wiley, New York
5. McCulloch WS, Pitts WH (1943) A logical calculus of the ideas immanent in nervous activity. Bull Math Biophysics 5:115–133
6. Rosenblatt F (1958) The perceptron: A probabilistic model for information storage and organization in the brain. Psychol Rev 65:386–408
7. Minsky M, Papert S (1969) Perceptrons, expanded edition. An introduction to computational geometry by Marvin Minsky and Seymour A. Papert. MIT Press, Cambridge
8. Möller K, Paaß G (1994) Künstliche neuronale Netze: eine Bestandsaufnahme. KI—Künstliche Intelligenz 4:37–61
9. Rummelhart DE, Hinton GE, Williams RJ (1986) Learning representation by back-propagating errors. Nature 323:533–536
10. Hornik K, Stinchcombe M, White H (1989) Multilayer feedforward networks are universal approximators neural networks. Neural Networks 2:359–366
11. Dean J (2014) Big data, data mining, and machine learning. Value creation for business leaders and practitioneers. Wiley, Hoboken
12. Jones N (2014) The learning machines. Nature 502:146–148
13. Hopfield JJ (1982) Neural networks and physical systems with emergent collective computational abilities. Proc Natl Acad Sci 79:2554–2558
14. Tank DW, Hopfield JJ (1991) Kollektives Rechnen mit neuronenähnlichne Schaltkreisen. Spektrum der Wissenschaft Sonderheft 11:65
15. Serra R, Zanarini G (1990) Complex systems and cognitive processes. Springer, Berlin, p 78
16. Hinton GE, Anderson JA (1981) Parallel models of associative memory. Psychology Press, Hillsdale N.J.
17. Ritter H, Martinetz T, Schulten K (1991) Neuronale Netze. Addison-Wesley, Bonn
18. von der Malsburg C (1973) Self-organization of orientation. Sensitive cells in the striate cortex. Kybernetik 14:85–100
19. Kohonen T (1991) Self-organizing maps. Springer, Berlin
20. Sutton R, Barto A (1998) Reinforcement-learning: an introduction. A Bradford Book, Cambridge
21. Russell S, Norvig P (2004) Künstliche Intelligenz: Ein moderner Ansatz. Pearson Studium, München

22. Pearl J (1988) Probabilistic reasoning in intelligent systems: networks of plausible inference, 2nd edn. Morgan Kaufmann, San Francisco
23. Tversky A, Kahneman D (2000) Advances in prospect theory: cumulative representation of uncertainty. In: Kahneman D, Tversky A (eds) Choices, values and frames. Cambridge University Press, Cambridge, pp 44–66
24. Damasio AR (1995) Descartes' Irrtum. Fühlen, Denken und das menschliche Gehirn. List, München
25. Picard RW (1997) Affective computing. MIT Press, Cambridge
26. Onisawa T (2000) Soft computing technology in Kansei (emotional) information processing. In: Liu Z-Q, Miyamoto S (eds) Soft computing and human-centered machines. Springer, Berlin
27. Velasquez JD (1997) Modeling emotions and other motivations in synthetic agents. Amer Assoc Art Int, pp 10–15
28. Flohr H (1991) Brain processes and phenomenal consciousness. A new and specific hypothesis. Theory and Psychology 1:245–262
29. Mainzer K (2008) Organic computing and complex dynamical systems. Conceptual foundations and interdisciplinary perspectives. In: Würtz RP (ed) Organic computing. Springer, Berlin, pp 105–122

Robots Become Social

8

8.1 Humanoid Robots

With the increasing complexity and automation of technology, robots are becoming service providers for industrial society. The evolution of living organisms today inspires the construction of robotic systems for different purposes [1]. As the complexity and difficulty of the service task increases, the use of AI technology becomes unavoidable. And robots don't have to look like humans. Just as airplanes do not look like birds, there are also other adapted shapes depending on their function. So the question arises for what purpose humanoid robots should possess which properties and abilities.

Humanoid robots should be able to act directly in the human environment. In the human environment, the environment is adapted to human proportions. The design ranges from the width of the corridors and the height of a stair step to the positions of door handles. For non-human robots (e.g., on wheels and with other grippers instead of hands) large investments for environmental changes would have to be made. In addition, all tools that humans and robots should use together are adapted to human needs. Not to be underestimated is the experience that humanoid forms psychologically facilitate the emotional handling of robots.

© Springer-Verlag GmbH Germany, part of Springer Nature 2020   141
K. Mainzer, *Artificial intelligence – When
do machines take over?*, Technik im Fokus,
https://doi.org/10.1007/978-3-662-59717-0_8

Unlike firmly anchored industrial robots, humanoid robots can fall over when walking when the foot floats freely [2, 3]. If one foot with the entire sole were always placed on the ground, a humanoid robot could walk safely without falling over.

In order to decide whether the contact of the sole with the ground is maintained, the ZMP (zero moment point) point is determined, i.e. the point on the ground at which all horizontal forces experienced by the foot from the ground are equal to zero.

The robot can rely on this to ensure that its current inclination remains unchanged. The ZMP is always located on the area between the sole and the ground (Fig. 8.1).

When a person stands upright, his sole retains ground contact as long as the center of gravity projected onto the ground is within the stability area of the sole with the ZMP (Fig. 8.1). However, this is only a sufficient condition to prevent the case, but not a necessary one.

Advanced problems like walking on uneven ground, climbing stairs, walking with heavy objects or racing require complex running patterns with the ZMP. Mathematically, the relationship

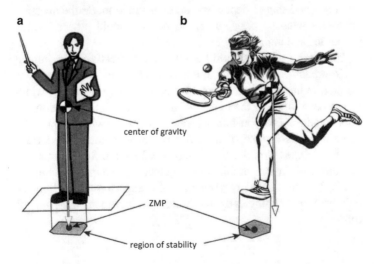

Fig. 8.1 Zero moment point (ZMP) and region of stability of a stationary (**a**) and a moving (**b**) human

between the trajectory of joint velocity and ZMP is given by a nonlinear differential equation system. The calculation of a joint velocity trajectory from a given trajectory of a ZMP is difficult [4]. Since ASIMO (2000), Honda has been using a process that produces a stable running pattern in real time.

Here the difference between technology and nature becomes clear. In order to achieve stable human locomotion, evolution did not require a high-performance computer capable of solving nonlinear equations of motion in real time in order to realize the movements according to the solution patterns. So, there was no need for "intelligent design" by an engineer. Mathematical models and corresponding computer programs are human inventions that solve the movement problem in their own way.

The dynamic model of forward kinematics shows how the next state of motion of a humanoid robot can be calculated for a given current state, forces and moments acting on the limbs, contacts with the environment, and other conditions.

The backward calculation of the forces and moments acting on the joints at a given nominal positioning is controlled in the dynamic model of inverse kinematics.

The underlying mathematical equations such as the Newton-Euler formula have long been known. Their technical implementation in real time only became possible since the high performance computers with efficient solution algorithms.

Humanoids robots have not only two legs and two arms. They are equipped with optical and acoustic sensors. In terms of space and battery life, there have been limitations with regard to the processors and sensors that can be used. Miniaturization of optical and acoustic functions is just as necessary as the development of distributed microprocessors for local signal processing.

Japan is the leader in humanoid robotics. The Japanese Ministry of Economy and Industry has been pursuing the Humanoid Robotics Project (HRP) since 1998 [5]. A humanoid robot should then be able to move freely in a normal environment, overcome stairs and obstacles, search for paths independently, remain mobile after a fall, operate doors independently and carry out work resting on one arm. In principle, a humanoid robot could then function like a human being.

After all, a robot should be able to independently perform motor tasks that any human being can perform. This requires three-dimensional optical sensors that perceive the condition, position, and direction of an object, a hand that can perform this task, and force sensors to detect the condition of the manipulator hand when gripping an object and planning the work steps.

The goal for 2020 would be a humanoid robot that shares the living space with humans and cooperates with them. The achievement of this objective would be the achievement of the final objective of the HRP. In this case, the humanoid robot should not injure humans or damage the environment. The safety and strength needed for movement and work should be equally guaranteed. Only then is a service robot available for humans, which, in principle, can be used in any household.

8.2   Cognitive and Social Robots

In order to reach the final stage of HRP, living together with humans, robots must be able to form a picture of humans in order to become sufficiently sensitive. This requires cognitive abilities. The three stages of the functionalist, connectionist, and action-oriented approach can be distinguished and will now be examined.

The basic assumption of functionalism is that there is an internal cognitive structure in living beings as in corresponding robots, which represents objects of the external world with their properties, relations and functions among themselves via symbols.

One also speaks of functionalism because the processes of the outside world are assumed to be isomorphic in the functions of a symbolic model. Just as a geometric vector or state space represents the motion sequences of physics, such models would represent the environment of a robot.

Background information
The functionalist approach goes back to the early cognitivist psychology of the 1950s by A. Newell and H. Simon [6]. The symbols are processed in a formal language (e.g., computer program) according to rules that establish logical relationships between the representations of the outside world, allow conclusions to be drawn and thus allow knowledge to be generated [7].

According to the cognitivistic approach, rule processing is independent of a biological organism or robot body. In principle, all higher cognitive abilities such as object recognition, image interpretation, problem solving, speech comprehension and awareness could be reduced to symbolic calculation processes. Consequently also biological abilities like e.g. consciousness should be transferable to technical systems.

The cognitivist-functionalist approach has proved its worth for limited applications, but it has fundamental practical and theoretical limitations. A robot of this kind requires a complete symbolic representation of the outside world, which must be constantly adjusted as the robot's position changes. Relations such as ON(TABLE,BALL), ON(TABLE,CUP), BEHIND(CUP,BALL) etc., which represent the relation of a ball and a cup on a table relative to a robot, change as the robot moves around the table.

People, on the other hand, do not need symbolic representation and no symbolic updating of changing situations. They interact sensory-physically with their environment. Rational thoughts with internal symbolic representation do not guarantee rational action, as simple everyday situations already show. Thus, we avoid a sudden traffic obstruction due to lightning-fast physical signals and interactions, without resorting to symbolic representations and logical derivations.

In cognitive science, we therefore distinguish between formal and physical action [8]. Chess is a formal game with complete symbolic representation, precise game positions, and formal operations. Soccer is a non-formal game with skills that depend on physical interactions without full representation of situations and operations. There are rules to the game. But because of the physical action, situations are never exactly identical and therefore cannot be reproduced at will (in contrast to chess).

The connectionist approach therefore emphasizes that meaning is not carried by symbols, but results from the interaction between different communicating units of a complex network. This formation or emergence of meanings and patterns of action is made possible by the self-organizing dynamics of neural networks (see Sect. 7.2) [9].

However, both the cognitivistic and the connectionistic approach can, in principle, disregard the environment of the systems and only describe symbolic representation or neuronal dynamics.

In contrast, the action-oriented approach focuses on embedding the robot body in its environment. In particular, simple organisms of nature such as bacteria suggest to build behavior-controlled artefacts that are able to adapt to changing environments.

But also here the demand would be one-sided, to favor only behavior-based robotics and to exclude symbolic representations and models of the world.

Background information

It is true to say that human cognitive performance takes into account functionalist, connectionist and behavioral aspects.

It is therefore correct to assume, as in humans, a humanoid robot's own corporeality (embodiment). Then these machines operate with their robot body in a physical environment and establish a causal relationship to it. They each have their own experiences with their bodies in this environment and should be able to build their own internal symbolic representations and systems of meaning [10, 11].

How can such robots independently assess changing situations? Physical experiences of the robot begin with perceptions about sensor data of the environment. They are stored in a relational database of the robot as its memory. The relations of the outside world objects form causal networks with each other, on which the robot orients its actions. In this context, a distinction is made between events, persons, places, situations, and objects of daily use. Possible scenarios and situations are described with propositions of a formal logic of 1st order (Fig. 8.2).

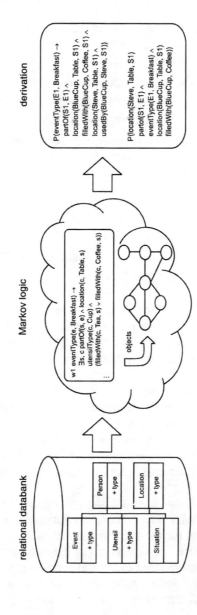

Fig. 8.2 Markov logic of a houshold robot

Example

If, for example, an event is of the type "breakfast", then kitchen utensils like cups are arranged in a certain way on a table and filled with tea or coffee:

eventType(e,Breakfast) → ∃s, c(partOf(s, e) ∧ location(c,Table,s)

∧ utensilType(c,Cup) ∧ (filledWith(c,Tea,s)

∨ filledWith (c,Coffee,s)))

Possible events depend on conditions that are associated with conditional probabilities in concrete situations. If we are talking about Steve's breakfast, there is a certain probability that he will use a blue cup for the coffee:

P(eventType(E1,Breakfast) | partOf(S1,E1)

∧ location(BlueCup,Table,S1)

∧ filledWith(BlueCup,Tea,S1) ∧ location(Steve,Table,S1)

∧ usedBy(BlueCup,Steve,S1))

The probability distribution of such situations is described in a Markov logic [12]. From this probability estimations of situations can be deduced [13, 14], at which a robot can orientate itself, if it is to prepare the breakfast for someone and to look for dishes in the kitchen.

The complex causal network of possible robot actions can be deduced from a Bayesian network of conditional probabilities. (see Sect. 7.2; [15, 16]). This does not in any way imply that human domestic helpers adhere to Bayesian networks in their actions. But with this combination of logic, probability, and sensory-physical interaction similar goals are realized as with humans.

Architecture of robot control is the name given to the arrangement of modules with their connections through which the reactions and actions of the robot are implemented. In a symbolically oriented architecture, the details of the hardware are abstracted and cognition is represented as symbol processing in the model. On the other hand, behavior-based architectures are based on an action-centered understanding of cognition. Physicality with all physical details, situatedness by the environment, and high adaptability play an important role. Behavior-based controls ensure that the robot responds quickly to environmental changes by processing stimuli perceived by sensors [17].

With symbolic processing, sensor inputs are first interpreted in an environmental model. Then, a plan for the action to be performed by actuators (e.g. wheels, feet, legs, arms, hands, grippers) is defined. This plan compares different goals as optimally as possible. The behavior-based approach does without sequential programming. Instead, as in a living organism, parallel processes must be coordinated.

Behavioral architectures [18] are found more in simple mobile robots, while symbolically oriented architectures are realized in cognitive systems with symbolic knowledge representation. Like humans, humanoid robots should also have both properties.

Humanoid robots are hybrid systems with symbolic knowledge representation and behavior-based action that take into account sensory-motor physicality and changes in environmental situations.

Hybrid systems are controlled in a hierarchically layered approach:

Complex behaviors at a higher level control one or more behaviors at an underlying level. A complex behavior is therefore composed of an ensemble of simpler behaviors.

In nature, this hierarchy often corresponds to the development of a living being in phylogeny.

Figure 8.3 shows the hybrid architecture of a humanoid robot with individual modules for perception, cognition, and action [19]. Cognition is divided into many sub-modules. Sensory data are thus interpreted, evaluated for linguistic representation, conceptual and situational knowledge in order to realize actions with sensor-motor skills. In this case, symbolic and sequential action planning is possible, but also quick reaction largely without the involvement of symbolic-cognitive instances.

Humanoid intelligence and adaptation, however, will only be able to develop if the artefacts not only have a body that is adapted and adaptable to their tasks, but can also react to the

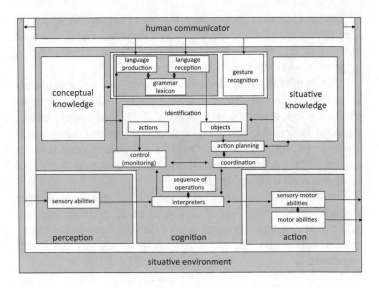

Fig. 8.3 Architecture of a humanoid robot with behavior-based and symbolic-cognitive modules

situation and largely autonomously. Since intelligence in living organisms such as humans develops and changes body-dependently in the course of a person's life, a growing body with highly flexible actuators will also become necessary. This requires cooperation with scientific disciplines that until recently seemed completely meaningless to the engineering sciences: cognitive science and brain research, ystems Bbiology and synthetic biology, nano and materials sciences.

8.3 Swarm Intelligence of Robots

In evolution, intelligent behaviour is by no means limited to individual organisms. Sociobiology regards populations as superorganisms capable of collective performance [20]. The corresponding abilities are often not completely programmed in the individual organisms and cannot be realized by them alone. An example is the swarm intelligence of insects, which can be seen in termite structures and ant trails [21]. Human societies with extrasomatic information storage and communication systems are also developing collective intelligence which only shows itself in their institutions.

Collective patterns and cluster formations can also be observed in populations of simple robots without having been programmed beforehand [22]:

Example

An example are simple robots that are only programmed to push small obstacles in front of them [23]. If the friction during pushing exceeds a threshold value, the robot vehicle turns in another direction.

One example are small insect-like robots that push tea lights together on a smooth surface. If a group of such robots encounters a randomly distributed number of equal obstacles, they will have pushed clusters of obstacles after a certain time.

There was no communication between the robots. The self-organization of the patterns is based solely on physical constraints of the collective interactions: The growth probability of a cluster therefore increases with increasing size, since the robots turn off when they bump into each other and leave their obstacles at the pile of obstacles. In this case there is still no pre-rational intelligence of the collective.

But if a collective of robots should have learned this behavior as useful (learning by doing) to repeat in future situations, collective intelligence would be realized.

Robot populations as service providers could find concrete application in road traffic with driverless transport systems or forklifts, which communicate independently about their behavior in certain traffic and order situations. Increasingly, different robot types such as driving and flying robots (e.g., for military missions or space exploration) will interact with each other [24].

R. A. Brooks of MIT generally calls for a behavioral AI which is based on artificial social intelligence in robot populations [25].

Social interaction and coordination of common actions in changing situations is an extremely successful form of intelligence that has evolved over evolution.

Even simple robots, like simple organisms of evolution, could generate collective achievements. In management, one speaks of social intelligence as a soft skill that should now also be considered by robot populations.

A first experimental field for such robot populations are the different varieties of robot soccer [26]. There are currently four categories of games. Only in the intended HuroSot (Humanoid Robot World Cup Soccer Tournament) class do humanoid soccer robots play on two legs with a size of approx. 40 cm. The MiroSot (Micro Robot World Cup Soccer Tournament) is a wheel-driven robot with an edge length of approx. 7,5 cm, whose teams are steered by central control computers. Somewhat smaller are the robots of the NaroSot (Nano Robot Cup Soccer Tournament) system. The KheperaSot (Khepera Robot Soccer Tournament) system also works with a central control computer.

Example

The current equipment of a robot team consists of three mobile robot systems:

- a central control computer,
- a (wireless) telecommunication system as a connection to the robots,
- an image processing system.

The central control computer calculates a game strategy with the next actions of the robots on the basis of the transmitted image data of the playing field. In addition to a drive mechanism, circuits for steering the drive mechanism and sensors, a play robot has a small computer that processes the sensor data and operation commands of the control computer.

The condition of a robot depends on the information where it is on the playing field, whether it is in possession of the ball and whether there is an obstacle to performing a play action.

Typical programmed behavioral patterns of a robot are "drive" to a specific position and "goal shot" if the connection between the standing position and the goal is unobstructed. To "intercept a ball" the path of a ball must be calculated from a previous and current position and the interception point must be determined.

No human player, of course, makes difficult geometric-mechanical calculations. Very few of these media stars would probably be able to do so at all. In any case, no one could do it in the speed, in order to be able to react afterwards also still lightning-fast.

This shows that the same performances of robots with mathematical models and high computing intensity are solved differently than by their biological colleagues. Similar to chess players, important soccer players use pattern recognition for match situations, which they compare fault-tolerantly and flexibly on the basis of their experience. In any case, this soccer knowledge is not stored in a rule-based way, but is available in a procedural way. In contrast to the chess player, the soccer player also has motor behavior patterns that have been trained prototypically. The communication system works with all non-technical forms of human message transmission, whereby body language with gestures and facial expressions plays a predominant role.

Robot soccer is an example of combined applications of machine learning (see Sect. 7.2). P. Stone speaks of "layered learning" in this context [27, 28]. A task to be solved is divided into several layers in order to solve the respective partial problems on these layers by appropriate learning algorithms (Fig. 8.4). The order of the applications of these algorithms is not rigid (e.g., from "top" to "bottom"), but results from the situation.

In addition, human players are subject to psychological factors such as motivation, mental strength, and moral. Again, emotional intelligence is required. Thus even outstanding teams can be completely demoralized and physically collapse by an unexpected course of play, while others keep their nerve to outgrow themselves in threatening situations with a view into the abyss.

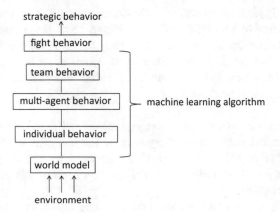

Fig. 8.4 Hierarchy of behavioral layers in robot soccer

Team and social behavior is of paramount importance in understanding human intelligence and performance. Behavior in sport is only a less complex and manageable field of experimentation than in business and society.

Thus, companies are also systems of people with feelings and awareness. In social groups, global opinion trends arise on the one hand through the collective interaction of their communicating members. On the other hand, global trends affect the members of the group, influence their microbehavior, and thus reinforce or slow down global system dynamics. Such feedback loops between micro and macro dynamics of a system enable learning effects in the company, such as anticyclical behavior, in order to counteract harmful trends. Digital models of production and organizational processes also serve this purpose.

References

1. Mainzer K (2010) Leben als Maschine? Von der Systembiologie zur Robotik und künstlichen Intelligenz. Mentis, Paderborn
2. Kajita S (ed) (2007) Humanoide Roboter. Theorie und Technik des Künstlichen Menschen. Aka, Berlin
3. Ulbrich H, Buschmann T, Lohmeier S (2006) Development of the humanoid robot LOLA. J Appl Mech Mater 5(6):529–539

4. Murray RM, Li Z, Sastry SS (1994) A mathematical introduction to robot manipulation. CRC Press, Boca Raton
5. Isozumi Akaike, Hirata Kaneko, Kajita Hiruka (2004) Development of humanoid robot HRP-2. J RSJ 22(8):1004–1012
6. Newell A, Simon HA (1972) Human problem solving. Prentice Hall, Englewood Cliffs
7. Siegert H, Norvig P (1996) Robotik: Programmierung intelligenter Roboter. Springer, Berlin
8. Valera F, Thompson E, Rosch E (1991) The embodied mind. Cognitive science and human experience. MIT Press, Cambridge
9. Marcus G (2003) The algebraic mind: integrating connectionism and cognitive science. MIT Press, Cambridge
10. Pfeifer R, Scheier C (2001) Understanding intelligence. A Bradford Book, Cambridge
11. Mainzer K (2009) From embodied mind to embodied robotics: humanities and system theoretical aspects. J Physiol (Paris) 103:296–304
12. Domingos P, Richardson M (2004) Markov logic: a unifying framework for statistical relational learning. In: Proceedings of the ICML Workshop on Statistical Relational Learning and Its Connections to Other Fields, S 49–54
13. Koerding KP, Wolpert D (2006) Bayesian decision theory in sensomotor control. Trends in Cogn Sci 10:319–329
14. Thurn S, Burgard W, Fox D (2005) Probabilistic robotics. MIT Press, Cambridge
15. Pearl J (2000) Causality, models, reasoning, and inference. Cambridge University Press, Cambridge
16. Glymour C, Scheines R, Spirtes P, Kelley K (1987) Discovering causal structures. Artificial intelligence, philosophy of science, and statistical modeling. Academic Press, Orlando
17. Braitenberg V (1986) Künstliche Wesen. Verhalten kybernetischer Vehikel. Vieweg + Teubner, Braunschweig
18. Arkin R (1998) Behavior-based robotics. A Bradford Book, Cambridge
19. Knoll A, Christaller T (2003) Robotik. Fischer Taschenbuch, Frankfurt, S 82 (nach Abb. 17)
20. Wilson EO (2000) Sociobiology: the new synthesis, 25th Anniversary Edition. Belknap Press, Cambridge
21. Wilson EO (1971) The insect societies. Belknap Press, Cambridge
22. Balch T, Parker L (eds) (2002) Robot teams: from diversity to polymorphism. A K Peters/CRC Press, Wellesley
23. Mataric M (1993) Designing emergent behavior: from local interaction to collective intelligence. In: From Animals to Animates 2 2nd Intern. Conference on Simulation of Adaptive Behavior, S 432–441
24. Mataric M, Sukhatme G, Ostergaard E (2003) Multi-robot task allocation in uncertain environments. Autonomous Robots 14(2–3):253–261
25. Brooks RA (2005) Menschmaschinen. Campus Sachbuch, Frankfurt

26. Dautenhahn K (1995) Getting to know each other—articial social intel-
    ligence for autonomous robots. Robotics and Autonomous Systems
    16:333–356
27. Stone P (2000) Layered learning in multiagent systems. A winning
    approach to robotic soccer. A Bradford Book, Cambridge
28. Leottau DL, Ruiz-del-Solar J, MacAlpine P, Stone P (2016) A study of
    layered learning strategies applied to individual behaviors in robot soc-
    cer. In: Almeida L, Ji J, Steinbauer G, Luke S (eds) RoboCup-2015:
    robot soccer world cup XIX, lecture notes in artificial intelligence.
    Springer, Berlin

Infrastructures Become Intelligent

9

9.1 Internet of Things and Big Data

The nervous system of human civilization is now the Internet. Up to now, the Internet has only been a ("stupid") database with signs and images whose meaning emerges in the user's mind. In order to cope with the complexity of the data, the network must learn to recognize and understand meanings independently. This is already achieved by semantic networks that are equipped with expandable background information (ontologies, concepts, relation, facts) and logical reasoning rules in order to independently supplement incomplete knowledge and draw conclusions. For example, people can be identified, although the data entered directly only partially describe the person. Here again it becomes apparent that semantics and understanding of meanings do not depend on human consciousness.

With Facebook and Twitter, we are entering a new dimension of data clusters. Their information and communication infrastructures create social networks among millions of users, influencing and changing society worldwide [1]. Facebook was created as a social network of universities (Harvard 2004). Social and personal data are always online. Data is by no means just text, but also images and sound documents.

Complex patterns and clusters are created in networks by locally active nodes. If people are influenced by the activity of their network neighbors,

© Springer-Verlag GmbH Germany, part of Springer Nature 2020   157
K. Mainzer, *Artificial intelligence – When do machines take over?*, Technik im Fokus,
https://doi.org/10.1007/978-3-662-59717-0_9

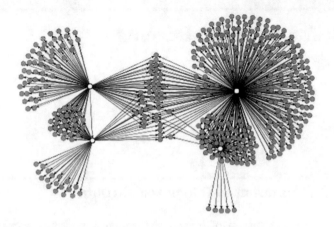

Fig. 9.1 Self-organization in a product network [2]

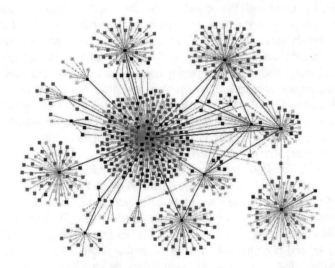

Fig. 9.2 Self-organization of an epidemic [3]

the adaptation to a new product or innovation can spread cascadingly in the network (Fig. 9.1; [2]). The spread of an epidemic disease (e.g., tuberculosis) is also a form of cascading pattern formation in the network (Fig. 9.2; [3]).

The similarity between biological and social patterns leads to interdisciplinary research questions. The local activity and mutual influence of the net nodes (whether customers or patients) can in principle be described by diffusion reaction equations. Their solutions correspond to patterns and cluster formations. If the parameter spaces of these equations are known, the possible cluster formations can be systematically calculated.

Self-organizing disciplines, applications, and stakeholders achieve offer and demand for new services and integrated solutions. But, while the classic Internet only supports communication between people in global computer networks, sensor technology opens up new possibilities for the future. A new dimension of communication: commodities, products, goods, and objects of all kinds can be equipped with sensors to exchange messages and signals. The Internet of persons transforms into the Internet of things:

▶ In the Internet of Things, physical objects of all kinds are equipped with sensors (e.g., RFID chips) to communicate with each other. This enables automation and self-organization of technical and social systems (e.g., factories, companies, organizations).

Hidden RFID and sensor technology creates the Internet of things that can communicate with each other and with people. For the Internet of Services, offers and technologies in the area of online commerce or online services and the media industry will be comprehensively expanded [4].

▶ Big data refers to the amount of data generated and processed on the Internet of Things. Not only structured data (e.g. digitized documents, e-mails) are recorded, but also unstructured data from sensors, which are generated by signals in the Internet of Things. The growing variety and complexity of services and possibilities in the network leads to an exponential data explosion. From petabytes (peta $=10^{15}$) up, an amount of data is called Big Data.

In the digital world, according to current estimates, the global volume of data doubles every two years. Under the term "Big

Data" experts summarize two aspects: on the one hand the ever faster growing mountains of data, on the other hand IT solutions and management systems with which scientific institutions and companies can evaluate, analyze, and derive knowledge from data. The industry that has developed around the collection, processing, and use of data is one in which corporations such as Google, Facebook, and Amazon are just the best-known representatives. Thousands of other companies thrive on generating, linking, and reselling information—a gigantic market. Big Data technology provides management with a significantly improved basis for time—critical decisions under increasing complexity [5].

Big Data refers to data sets whose size and complexity (petabyte range) is not possible due to classical databases and algorithms for collecting, managing, and processing data at manageable costs and in the foreseeable future. Three trends need to be integrated:

- massive growth of transaction data volumes (big transaction data),
- explosive increase of interaction data (big interaction data): e.g. social media, sensor technology, GPS, call logs,
- new highly scalable and distributed software (Big Data processing): e.g., Hadoop (Java) and MapReduce (Google).

Example

An example is the MapReduce algorithm which uses the functions "map" and "reduce" from functional programming and handles large amounts of data by parallel conputation [6]. In order to explain the principle, we will consider a simplified example: It is to be determined in an extensive data set how often which words occur. First, the entire text is divided into data packages, for which the frequency of words in the individual subpackages is computed in parallel using the map function. These partial results are collected in intermediate result lists. Using the reduce function, the intermediate result lists are merged and the frequencies for the entire text are computed.

Hadoop is a framework written in Java for distributed software that uses the MapReduce algorithm. It is used by, e.g., Facebook, AOL, IBM, and Yahoo. The credit card company Visa thus reduced the processing time for evaluations of 73 billion transactions from one month to approx. 13 min.

Big Data initially means huge amounts of data [7]: Google handles 24 petabytes a day, YouTube has 800 million monthly users, Twitter registers 400 million tweets a day. Data is analog and digital. They concern books, pictures, e-mails, photographs, television, radio, but also data from sensors and navigation systems. They are structured and unstructured, often not exact, but exist in masses. By using fast algorithms, they should be transformed into useful information. This means the discovery of new connections, correlations, and the derivation of future prognoses.

However, forecasts are not necessarily extrapolated on the basis of representative samples using conventional statistical methods. Big data algorithms evaluate all data in a data set, however large, diverse and unstructured they may be. What is new about this evaluation is that the contents and meanings of the data records do not have to be known in order to be able to derive information.

This is possible by so-called metadata [8]. What this means is that we do not need to know what someone is talking about on the phone, but the movement pattern of their mobile phone is decisive. A precise movement pattern of the mobile phone user can be determined over a certain period of time from a data retention memory, since the local radio cells are switched on with every automatic e-mail query and another use. In Germany, there are about 113 million mobile phone connections whose sensors and signals function like a measuring device.

The data in an e-mail refers to the text of the content. Metadata of the e-mail are, e.g., sender, recipient, and the time of sending. In the immersion project of the Media Lab of the MIT (Massachusetts Institute of Technology), graphs are automatically drawn from such metadata. In an earlier experiment at MIT, motion patterns of 100 people had been determined over a recording period of 450,000 h. This made it possible to determine who met whom and how often at certain locations. Places were grouped as workplace, home and others. On the basis of corresponding patterns of metadata, friendships could be predicted with a probability of 90%.

Often, however, predictions can only be derived from metadata if the correct contexts are known. Today, however, there are databases and background information on the Internet with which the meanings can be made accessible. In principle, this development of meanings works like a Semantic Web. The discovery of an American bioinformatician who used metadata alone to determine the name of an anonymous donor of human genetic material was spectacular. Metadata related to the age of the donor and the name of the American state in which the donation was made. The bioinformatician limited the search by combining place and age and used an online search engine in which families entered the genetic code for genealogical research. In the process, family members of the wanted persons emerged, whose data she combined with demographic tables, in order to finally find what they were looking for.

Even in medicine, the mass evaluation of signals leads to surprisingly fast predictions. Thus, the outbreak of an influenza epidemic could be predicted weeks earlier than was usually possible with data collection and statistical evaluations by health authorities [9, 10]. One had simply evaluated the behavior of people on the basis of billions of data in social networks and significant correlations which indicate the outbreak of the epidemic on the basis of previous experience.

In view of such examples, P. Norvig (Google) spoke about the "Unreasonable Effectiveness of Data" with allusion to the Nobel laureate in physics E. P. Wigner, who had highlighted "The Unreasonable Effectiveness of Data" [11]. However, current medicine also provides counter-examples for the effectiveness of Big Data if you don't know the reasons and causes. None other than S. Jobs, who became a symbol of more effective and smart computer technology, died of cancer, although he was able to use all the computing capacity and Big Data analysis available at that time with his money. DNA sequencing still cost a lot of computing power and money back then. Jobs had his cancer cells sequenced at short intervals in order to be able to continuously adapt the appropriate drug treatment.

▶   **Important** As long as the causal causes of a data
        correlation (e.g. biochemical basic laws and cellular
        mechanism in cancer) are not known and not under-
        stood, the mass evaluation of data and the calcula-
        tion of correlations only help to a limited extent:

        "Correlation is no causation!"

Thus, predicative modeling is the central goal of Big Data mining as part of data science. Algorithms of machine learning are
used for this purpose. They are modelled on the human brain
from the neurosciences and AI research with pattern formation
and clustering. But, they are also based on statistical and database methods (Fig. 9.3; [12]). With Big Data, a collective intelligence is emerging that is not attached to individual organisms,
brains, or computers. Rather, it is part of the global information
and communication network of the Internet of Things.

Boltzmann machines (BM) are neural networks with stochastic learning algorithms (G. Hinton 1980), which we have already
met in Sect. 7.2. In the Internet of Things they realize today deep
learning, e.g., through recommendation systems for millions of
products and customers in social networks.

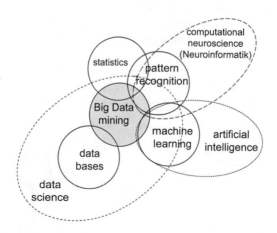

Fig. 9.3 Big Data and Artificial Intelligence [12]

Example

A limited Boltzmann machine (BBM) consists of two neural layers with stochastic units (Fig. 9.4; [13]). The visible neurons v_i represent recommendations with discrete values between 1 and Q. The q-th value v_i^q is activated with a probability depending on the input. The connection between the k-th hidden neuron and the q-value of the i-the visible neurons is weighted with w_{ki}^q.

The weights of a BBM are learnt by maximizing the probability $p(V)$ of the visible units ("recommendations") V. It depends on the distribution of the "computing energy" $E(V, h)$ in the entire network. The network is "cooled down" like a molecular liquid to a state of equilibrium, which is linked to the recommendations. The gradient descent of this cooling corresponds to a stochastic learning rule.

In Sect. 7.2, multi-layer neuron networks were illustrated that recognize faces. In the 1980s, that was just mathematical theory. The computers and memory capacities were too small at that time to realize these learning algorithms technically. In the meantime, brain research has empirically proven that these models at least approximate reality. In the age of Big Data, however, these models also become technically feasible. We are talking about "deep learning": learning to learn machines.

The model is not mathematically new. As in the models from the 1980s, neural networks in deep learning are arranged into layers that use increasingly complex features, such as, e.g., to recognize the content of an image. This allows even large amounts of data to be divided into categories. In the "Google

Fig. 9.4 Deep learning in the Internet of Things [12]

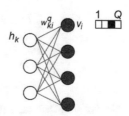

Brain" (Mount View CA 2014), approximately 1 million neurons and 1 billion compounds (synapses) are simulated. Big data technology makes neural networks with multiple intermediate steps possible, which were only theoretically conceivable in the 1980s.

9.2   From Autonomous Vehicles to Intelligent Transport Systems

About 50 years ago, people were on the moon and needed less computer power than today's laptops. It will start again soon, but with an intelligent and autonomous robot vehicle. Target is the Apollo 17 mission landing site. The project is called "Part-Time-Scientists" and the competition is announced by Google: Google Lunar XPRIZE [14]. The German carmaker Audi will also be there. This autonomous vehicle is to be made possible by proven Audi technology—through its all-wheel drive and lightweight construction, an electric motor and artificial intelligence of autonomous driving [15].

The Prize money is $30 million. However, the mobile robot must be brought to the moon self-financed and drive 500 m there. A rocket is planned to carry the Audi lunar quattro (Fig. 9.5) into space. Other cooperation partners are NVIDIA, TU Berlin, the Austrian Space Forum (OeWF), and the German Aerospace Center (DLR).

The autonomous vehicle has a maximum speed of 3.6 km/h. Its chassis is equipped with double wishbones for all four wheels, each of which can be rotated 360 degrees. The energy is supplied by solar cells that drive four wheel hub motors. A swivelling solar panel of approx. 30 cm^2 size captures the light of the sun and converts it into electricity. In addition, there is a lithium-ion battery which is housed in the chassis. Their energy should make it possible to cover the distance required by the competition. When the sun shines, the temperature on the lunar surface rises up to 120 °C.

For such extreme conditions, the moon rover is largely made of high-strength aluminium. But its current weight of 35 kg is still too high to fly to the moon. The weight is to be further

Fig. 9.5   Autonomous robot mobile Audi lunar quattro [15]

reduced through the use of magnesium and design changes. Less weight means less fuel for the Lunar Module and less costs for the launch vehicle.

What is decisive, however, is the ability to find one's way around obstacles on one's own. A moving head at the front of the vehicle carries two cameras that record detailed 3D images. A third camera is used to examine materials and provides extremely high-resolution panoramas.

The small project team of ten employees combines the competence of a highly qualified automotive industry with ICT technology. Just like 50 years ago, the flight to the moon will become a laboratory for future technology on Earth—in this case the production of autonomous and intelligent vehicles.

Car manufacturer Ford also cooperates with Silicon Valley and wants to produce autonomous vehicles connected with 3D printing and wearables. Over the next years, driver assistance systems [16] are to be further developed into autonomous vehicles. As early as the 1990s, the Eureka Prometheus project was talking about (e.g., Munich University of the Federal Armed Forces) of "intelligent automobiles" and "intelligent roads" of the future [17]. All automobile models are to be equipped with the pre-collision assistant, including pedestrian recognition technology. Finally, 3D printing processes will be used to produce vehicles from lightweight composite materials and to extend networking to smartwatches and other wearables.

Autonomous vehicle steering has its origins in military research. The DARPA (Defence Advanced Research Projects Agengy) has been working on the development of autonomous military vehicles since the 1970s. According to a US Congress decision, one third of all US military vehicles should be able to drive without a driver by 2015. The civilian use of autonomous vehicles was early adopted by Google is being pushed forward. Projects of German universities such as TU Braunschweig, FU Berlin and TU Munich are also worth mentioning.

Google has already been granted a patent for its technology and has mastered several hundred thousand kilometres accident-free with its prototype. The central component of the Google vehicle is the Velodyne HDL-64E LiDAR sensor unit on the roof [18]. Through rotation, this sensor generates a 3D model of the environment. Laser pulses are used to measure distance and speed. High-resolution images provide a real-time picture of the surrounding traffic. The data acquisition amounts to one gigabyte per second. It enables an environment map to be stored and can be extended again and again with further observation data of the autonomous vehicle (Fig. 9.6).

Flowing traffic is detected by radio sensors in the front and rear bumpers. Traffic signs and traffic lights are registered by a camera. These images are processed by a software into environmental information for the control unit. Vehicle movements are determined by sensors in the tires, a GPS module and inertial sensors to calculate route and speed in real time. Together with the control unit, critical situations can thus be avoided. In order to gain experience and improve the maps of the road network, Google engineers travel the routes several times before the vehicle is allowed to drive autonomously. The challenges of situation analysis, decision making, and behavior increase with higher driving speed.

Autonomous reactions in different situations without human intervention are a major challenge for AI research. Decision algorithms can best be improved in real road traffic. Analogously, a human driver improves his skills through driving

Fig. 9.6 Environment image of the autonomous Google vehicle [18]

practice. The robot car from Google already passed the Turing test in so far as journalists cannot tell the difference between his demonstration and a vehicle driven by humans. People will be the biggest mistake factor in the long run. It remains to be seen whether they want to do without the fun of driving.

However, error factors can also be companies that overestimate their possibilities. Tesla should have known that a few years ago its AI software was not yet able to clearly distinguish moving vehicles from their background. So, it happened that a partially autonomous vehicle of this company drove into a truck at an intersection, because the software confused the large bright area of the loading space with the background of the sky—with deadly outcome for the human passenger. Risk assessment with valid basic research belongs to the technical, legal and ethical safety standards of engineering science (cf. Sects. 11.1, 12.2).

We sum up: A self-propelled motor vehicle or robot car is a vehicle that can drive, control and park without a human driver.

Highly automated driving lies between assisted driving, in which the driver is supported by driver assistance systems, and autonomous driving, in which the vehicle drives automatically and without the influence of the driver.

In highly automated driving, the vehicle only partially has its own intelligence, which plans ahead and could take over the driving task at least in most situations. Man and machine work together.

9.3   From Cyberphysical Systems to Intelligent Infrastructures

Classical computer systems were characterized by a strict separation of the physical and virtual worlds. Mechatronics control systems, such as those installed in modern vehicles and aircraft and consisting of a large number of sensors and actuators, no longer correspond to this picture. These systems recognize their physical environment, process this information and can also influence the physical environment in a coordinated way [19]. The next step in the development of mechatronic systems is the "Cyberphysical Systems" (CPS), which are not only characterized by a strong coupling of the physical application model and the computer control model, but are also embedded in the work and everyday environment (e.g., integrated intelligent power supply systems) [20]. Through networked embedding in system environments, CPS systems go beyond isolated mechatronic systems.

▶ Cyberphysical Systems (CPS) consist of many networked components that independently coordinate among themselves for a common task. They are thus more than the sum of the many different smart small devices in ubiquitous computing, since they implement complete systems from many intelligent subsystems with integrating functions for specific goals and tasks (e.g., efficient power supply) [21]. This extends intelligent functions from the individual subsystems to the external environment of the overall system. Like the Internet, CBS become collective social systems which, in addition to information flows, integrate energy, material and metabolic flows (such as mechatronic systems and organisms).

Historically, CPS research originated in the field of "embedded systems" and mechatronics [22]. The embedding of information and communication systems in work and everyday environments led to new performance requirements such as fault tolerance, reliability, failure, and access security with simultaneous implementation in real time. However, weak points became increasingly noticeable when embedding corresponding control processes. Examples are automatic traffic systems for congestion avoidance and the harmonization of individual travel times with economically and ecologically efficient solutions [23]. Equally difficult was the supply of battery-powered electric cars via regenerative energy conversion systems such as solar cells or wind turbines. This also includes renewable energies, which are made available as a sufficiently reliable and cost-effective alternative or reserve energy for supply networks.

In these increasingly complex applications, high adaptability of system control and system architecture, dynamic process behavior, fast overcoming or repair of failures, expansion and enlargement of the system are required. The main obstacle to the realization of these requirements used to be the attempt to centrally control the entire system. The evaluation of global information simply took too long to be able to initiate appropriate control measures. For example, large transport systems are highly dynamic. Even if traffic jam messages were distributed every two minutes, they could not be evaluated quickly enough to adapt to the traffic situation. As a consequence, the navigation instruments in trucks calculate their individual alternative routes. Since, however, all devices use the same statistical algorithm, all vehicles are diverted to the same route to avoid traffic jams and thus increase chaos.

CPS therefore aim to adapt control processes and information flows to the physical processes of their applications [24], as evolution has achieved in the development of its organisms and populations. Top-down structures of software, which are imposed on physical processes "from above", are not a solution. Distributed control, bottom-up management for layered control structures, highly autonomous software processes and distributed learning strategies for agents are the benchmarks.

Example

One example is smart grids, which, in addition to conventional electricity transport, also allow data communication in order to meet the requirements for highly complex network operation. The trend is towards global and transnational network structures such as the Internet, in which combined heat and power plants for generating electricity from fossil primary energy are represented, as well as renewable sources with photovoltaic plants, wind power plants, and biogas plants. Consumers such as residential houses or office buildings can also be local power producers with photovoltaic systems, biogas, or fuel cells, who supply themselves or their environment with energy [25]. These housing estates implement the principle of local activity, whereby input from a domestic energy source is fed into the grid environment and contributes to global distribution patterns.

Smart grids with integrated communication systems thus implement a dynamically regulated energy supply [26]. They are an example for the development of large and complex real-time systems according to the principles of cyberphysical systems. The reserve energy to compensate for short-term load peaks or voltage drops is traditionally held centrally by large power plants. It is important to reallocate the total energy in the grid intelligently, flexibly, and according to demand. The main problem with switching to renewable energies lies in the large number of constraints that have to do with functional operation, safety, reliability, time availability, fault tolerance, and adaptability.

Cyberphysical systems with decentralized and bottom-up structures are therefore the answer to the increasing complexity of our supply and communication systems. Central to this is the organization of data streams that control the energy supply like in the nervous system of an organism. In the IT world, the term "cloud" is used when data is no longer stored in the home computer but in the network. The cloud is then a virtual network storage for big data. The network itself is ultimately a universal Turing machine in which the data processing of many computers is recorded.

According to Church's thesis, effective data processing can be realized in various ways if it is mathematically equivalent to a Turing machine. From cellular automata to neural networks and the Internet, network structures are created in nature and technology in which the elements of complex systems interact according to local rules. Locally active cells, neurons, transistors and network nodes generate complex patterns and structures that are linked to the collective performance of the system—from life functions in organisms through cognitive performance of the brain and swarm intelligence of populations to the organization of technical infrastructures such as the energy system. In order to master the computability of these systems, we need to know the mathematics of networks.

The first practical challenge of networking is the digitization of existing infrastructures. Historically, the infrastructures that exist today have been created separately and uncoordinated as completed systems. These include transport, energy, health, administration, and education. The Internet of Things leads to overlapping fields of application such as smart home, smart production, smart city, and smart region. The intelligent networking of previously separate domains makes new efficiency and growth potentials possible. However, new tasks of technical, economic, legal, regulatory, political, and social integration are also emerging.

▶ Intelligent networks and services are created by linking classical infrastructures and complementing artificial intelligence (i.e. autonomously operating and controlling functions and components). The intelligence of infrastructures and networks is therefore a capability that arises "vertically" within a domain (such as health and transport) and "horizontally" across domains (cf. Fig. 9.7; [27]).

Six domains for the supply infrastructures transport, energy, health, education, administration and information and communication technology (ICT) are distinguished in the Fig. 9.7. In its linking function, the ICT domain fulfils a fundamental prerequisite for the emergence of cross-domain services (e.g., connection of transport and energy supply or health with education and information systems). Since cyberphysical systems connect digital with analog functions (e.g., programs with sensors) (e.g. smart cars), they will act in intelligent networks as nodes at the interface of the digital and analog world.

9.4 Industry 4.0 and the Labor World of the Future

Big Data is closely related to social networks in everyday life and industry 4.0 as a trend of the labor world. Big Data uses not only structured business data of a company, but also unstructured data from social media, signals from sensors, and audio and video data.

▶ Industry 4.0 alludes to the preceding phases of industrialization. Industry 1.0 was the age of the steam engine. Industry 2.0 was Henry Ford's assembly line. The assembly line is nothing more than an algorithmization of the work process, which realizes a product step by step according to a fixed program through the division of labor and the use of people. In Industry 3.0, industrial robots intervene in the production process. However, they are fixed locally and always execute the same program for a specific sub-task. In Industry 4.0, the work process is integrated into the Internet of Things. Workpieces communicate with each other, with transport facilities and people involved, in order to organize the work process flexibly.

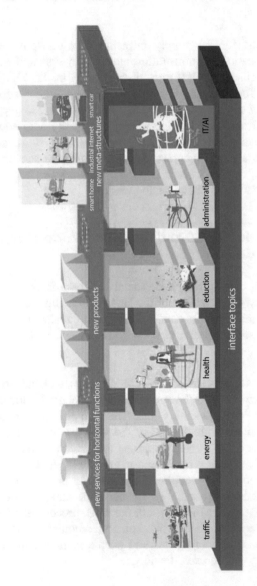

Fig. 9.7 Digitization of supply infrastructures in intelligent networks [27]

In Industry 4.0, products can be created individually at the desired time according to customer requirements. Technology, production, and market merge into a socio-technical system that organizes itself flexibly and adapts automatically to changing conditions. This is the vision of a cyberphysical system for industry [28]. For this purpose, machines and sensor data must be connected, captured, transported, analyzed and communicated with text documents. The Big Data technology used for this purpose is aimed at faster business processes and thus, it is hoped, faster and better decisions.

Industry 4.0, however, is not a completely new technology push, but has been prepared by various steps of intelligent problem solving:

Example

A first example is computer-aided design (CAD). CAD applications are supported by AI in the form of expert systems. Instead of a 2-dimensional drawing, a 3-dimensional virtual model of a production object can also be imported for construction purposes. Together with the respective material properties, CAD computer models support design and production in all technical applications from mechanical and electrical engineering to civil engineering and architecture.

Another example are CNC (computerized numerical control) machines. These machine tools are no longer controlled mechanically, but electronically by programs. The data are read from the CAD program of the production design into a CNC program in order to control the material production. In the meantime, both quality control and tool wear monitoring can be taken into account fully automatically in the manufacturing process.

An illustrative example should illustrate the development of the industrialization stages up to industry 4.0: The lathe was historically developed in pre-industrial times by wood turners and carpenters to turn workpieces (e.g., legs of chairs and tables). In the age of industrialization, they became metal-working lathes in which the production of the workpieces (e.g. gears, shafts) depended on the skill of the lathe operator. A CNC lathe has a computerized numerical control (Fig. 9.8). Computer-assisted information for machining is read into the memory of the control, in order to be applicable again and again for machining. The control data are entered numerically, i.e. in numerical codes, and are repeatedly reconciled during the production process.

Basic for CNC machines is CNC programming [29]. A distinction is made between programming directly at the machine, with a laptop or via corresponding networks:

▶ **Definition**
The most important addresses (letters) in a CNC program are:

T ... calls the corresponding tool (e.g. B. T0101 or T01, T = Tool)

S ... selection of spindle speed (two to six digits, e.g. B. S800 S = Speed)

M ... so-called modal functions (one to three digits, depending on the manufacturer, e.g. B M08 coolant supply ON M = Modal)

G ... displacement commands (one to three digits, depending on the manufacturer, e.g. B G00 straight tool movement at rapid traverse G = Go)

X, Y, Z, U, V, W, I, J, K, C are coordinates to which the tool moves.

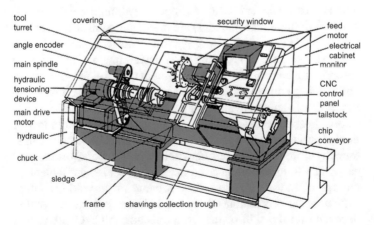

Fig. 9.8 Functional units of a CNC-lathe [29]

There are also parameters for lengths, angles, and other additional functions.

CNC blocks consist of the digits 0–9 with different meanings. Addresses must always be placed before blocks, since the controller does not recognize a block alone.

Example of a CNC program line:

```
G01 X135.5 Z7.2 F0.05 A150.;
```

The sentences have the following meaning:

G01   operation, straight travel of the tool carrier,
X135.5   go to absolute coordinates Z7.2,
F0.05   with a feed rate of 0.05 mm/rotation,
A150   at a 30 degree angle.

Addresses and records are combined into lines in which all commands to the controller are processed independently by the machine. The lines must be terminated at the end with a command character (e.g. B.;), otherwise the controller will not recognize the end of the line.

As part of Industry 4.0, CNC machines are networked with each other, communicate with workpieces via RFID chips and carry out measurements independently. Feeders and removal systems are also automated. The use of social and cognitive robots in the workplace is now also conceivable. This further relieves the operating personnel and improves productivity. However, qualified personnel are required for adjusting and setting up the machines.

Industry 4.0 makes new customer-oriented production possible: on-demand production or tailored production. In the past, only a few rich people could afford tailor-made suits, individually for their own needs. In industry 4.0, on-demand production is carried out according to the individual and personalized customer design. The individual and personalized production process can organize itself. We are also observing the trend towards decentralized and individual supply in energy systems. So, all along the line we are experiencing a departure from mass and standard production à la Henry Ford—from industry to nutrition to personalized medicine.

We are currently talking about a market of around 14 billion globally networked devices, a third of which are in the USA. By 2020, the number of networked devices will have doubled. One reason is the exponential development of sensor technology and computing capacity.

▶ According to Moore's law, the computing power of computers doubles every approx. 18 months—with simultaneous miniaturization and price reductions for the devices.

▶   **Important** The ICT world is thus driven by exponential growth laws:
        Exponentially   growing   computing   capacity (Moore's law)
        Exponentially growing number of sensors
        Exponentially growing masses of data etc. etc.

Companies must adapt their corporate structures and become intelligent problem solvers: ICT technology is increasingly reducing traditional material production to apps and software modules: Cameras become apps in smartphones. Google, a prime example of an exponentially growing IT company, is already building autonomous electric cars. Kodak as a mass producer of camera equipment has largely disappeared from the market, because everyone has these camera functions in tiny apps and sensors in their smartphones. What happens if this strategy is applied to the automotive industry and cheap 3D printers soon produce the material building blocks of an automobile? Then, it only depends on the data that has to be put into these 3D printers—and those who can handle this data.

IT companies are breaking into other domains everywhere. Conversely, IT companies will also have to adapt. An IT giant like Microsoft has so far produced industry 2.0 style software as mass and standard fabrication. In the world of industry 4.0, softwarc houses will have to address individual wishes and the needs of individual corporate customers. Corporations will no longer be able to set mass standards, but will be transformed into consulting firms that have to develop individual tools and tailor-made IT infrastructures with their customers. This also applies to energy companies. They are increasingly focusing on a decentralized market and are relying on individual advice to find the right solution. This leads to new business models such as buy and build. Trust can only be built through individualization and personalization. This is the necessary accompanying measure to make intelligent companies a reality.

Where is man? His trust is critical to the success of the industrial Internet and intelligent automation. Because there's skepticism, too. The reason is the cloud technology: If a medium-sized entrepreneur earns good money with his business model, then he will be careful not to place the corresponding data in the cloud—because of industrial espionage, but also because of the high investments that have to pay off. The previous safety technology is the Achilles' heel of Industry 4.0. Therefore, individual solutions will also have to be found in this case. You will need to think carefully about what data to put in the cloud to enable effective access to data by employees and customers. Especially sensitive company data does not belong in the cloud. In the given cases, industry 3.1 or 3.3 are also individually good solutions, which depend on the respective company profile.

The question of data security also arises from the employee's point of view: Automation is only possible because many sensors, cameras, photoelectric sensors, etc. permanently record masses of data. Who has access to this data, where is it stored how long for whom?

After all, it is about the labor market itself. Will industrial automation not lead to unemployment? Is artificial intelligence appearing here as a threat to humans? The development of intelligent factories will primarily increase the effectiveness of the industry. It will lead to the dismantling of routine and mechanical work—both manual and intellectual. However, this is not new and has accompanied the industrialization process since the 19th century. New jobs will be created for this. Customer service is of particular importance here, as communication with customers and the development of business models not only require a wide range of knowledge in business and management, but also flexibility, experience, and psychology in dealing with customers. There are also professions in the field of mechatronics and robotics.

AI-supported automation therefore does not produce unemployment, but lowers production costs and thus promotes the labor market for a wide range of qualified employees. This will enable a country with the appropriate qualifications to regain production in low-wage countries. Germany, which is already highly automated, has significantly lower unemployment than other European countries. Unemployment in these countries has other reasons and is related to the lack of reforms of the labor market. We will not only need highly qualified engineers with university degrees and do the rest machines. We will continue to need people's know-how in all areas.

In engine and plant construction engineers will have to be trained in mechanical engineering, electronics, and information technology. In the tradition, these were disciplines from outside the field. Engineers will work in teams with different specializations to solve complex industry 4.0 networked problems. Interdisciplinary cooperation skills are becoming an indispensable training requirement.

In the metalworking professions, there will continue to be the "lathe operator", but as a specialist for networked CNC lathes. The requirements will change in the process. The innovation cycles are already faster than our training cycles in many areas. In future, we must therefore consider what we are actually training people for. If we teach someone a certain computer program today, it will be outdated by the time they come into operation. That is why we need to train people's ability to familiarize themselves with new work processes and to adapt to new situations. In the future, it will be an absolute norm that a part of the employees will always be in training courses to prepare themselves for new processes:

▶   Automation and intelligent enterprises require life-long learning!

References

1. Easley D, Kleinberg J (2010) Networks, crowds, and markets. Reasoning about a highly connected world. Cambridge University Press, Cambridge
2. Leskovec J, Adamic L, Huberman B (2007) The dynamics of viral marketing. ACM Transactions of the Web 1(1):5
3. McKenzie A, Kashef I, Tillinghast JD, Krebs VE, Diem LA, Metchock B, Crisp T, McElroy PD (2007) Transmission network analysis to complement routine tuberculosis contact investigations. Am J Public Health 97(3):470–477
4. Mainzer K (2014) Die Berechnung der Welt. Von der Weltformel zu Big Data. Beck, München
5. BITKOM (ed) (2012) Big Data im Praxiseinsatz – Szenarien, Beispiele, Effekte. BITKOM, Berlin
6. Dean J, Ghemawat S (Google Labs) MapReduce: Simplified Data Processing on Large Clusters. http://research.google.com/archive/mapreduce.html
7. Mayer-Schönberger V, Cukier K (2013) Big data – a revolution that will transform how we live, work and think. Eamon Dolan/Mariner Books, London
8. Hambuch U (2008) Erfolgsfaktor Metadatenmanagement: Die Relevanz des Metadatenmanagements für die Datenqualität bei Business Intelligence. VDM Verlag Dr. Müller, Saarbrücken
9. Ginsburg J (2009) Detecting influenza epidemics using search engine query data. Nature 457:1012–1014
10. Dugas AF (2012) Google flu trends: correlations with emergency department influenza rates and crowding metrics. CID Advanced Access 8. https://doi.org/10.1093/cid/cir883
11. Halevy A, Novik P, Pereira F (2009) The unreasonable effectiveness of data. IEEE Intelligent Systems March/April, pp 8–12
12. Dean J (2014) Big data, data mining, and machine learning. Value creation for business leaders and practioners. Wiley, Hoboken, p 56 (after Fig. 4.1)
13. Salakhutdinov R, Hinton G (2007) Restricted Boltzmann Machines for collaborative filtering. In: Mnih, Mnih A (Hrsg) Proceedings of the ICML, pp 791–798
14. http://lunar.xprize.org/. Accessed: 30. Juli 2015
15. http://www.audi.de/de/brand/de/vorsprung_durch_technik/content/2015/06/mission-to-the-moon.html. Accessed: 30. Juli 2015
16. Stiller C (2007) Fahrerassistenzsysteme. Schwerpunktthemenheft der Zeitschrift it – Information Technology 49:1

17. Braess HH, Reichart G (1995) Prometheus: Vision des „intelligenten Automobils" auf „intelligenter Straße"? Versuch einer kritischen Würdigung. ATZ Automobiltechnische Zeitschrift 4:200–205
18. http://www.velodynelidar.com/lidar/products/manual/HDL-64E%20 Manual.pdf. Accessed: 30. Juli 2015
19. Hawkins W, Abdelzaher T (2005) Towards feasible region calculus: An end-to-end schedulability analysis of real-time multistage execution. IEEE Real-Time Systems Symposium. 1 pp 2–88, Miami Florida
20. Lee E (2008) Cyber-physical systems: Design challenges. In: Technical Report No. UCB/EECS-2008-8. University of California, Berkeley
21. Cyber-Physical Systems (2008) Program Announcements & Information. The National Science Foundation, 4201 Wilson Boulevard, Arlington, Virginia 22230, USA, 2008-09-30
22. Wayne W (2008) Computers as components: principles of embedded computing systems design. Morgan Kaufmann, Amsterdam
23. Wedde HE, Lehnhoff S, van Bonn B (2007) Highly dynamic and adaptive traffic congestion avoidance in real-time inspired by honey bee behavior. In: PEARL Workshop 2007, Informatik aktuell. Springer
24. Broy M (1993) Functional specification of time-sensitive communication systems. ACM Transactions on Software Engineering and Methodology 2:1–46
25. European Technology Platform Smart Grids. http://ec.europa.eu/ research/energy/pdf/smartgrids_en.pdf. Accessed 30. Juli 2015
26. Wedde HF, Lehnhoff S (2007) Dezentrale vernetzte Energiebewirtschaftung im Netz der Zukunft. In: Wirtschaftsinformatik 6
27. Informations- und Kommunikationstechnologien als Treiber für die Konvergenz Intelligenter Infrastrukturen und Netze (2014). Studie im Auftrag des Bundesministeriums für Wirtschaft und Energie (Projekt-Nr. 39/13). LMU-Forschungsverbund: Intelligente Infrastrukturen und Netze, Abb. 2, S. 20
28. acatech (Ed.) (2011) Cyber-Physical Systems. Innovationsmotor für Mobilität, Gesundheit, Energie und Produktion, Berlin
29. Falk D (2010) CNC-Kompendium PAL Drehen und Fräsen, Braunschweig. https://de.wikipedia.org/wiki/CNC-Drehmaschine. Accessed 30. Juli 15

From Natural and Artificial Intelligence to Superintelligence?

10

10.1 Neuromorphic Computers and Artificial Intelligence

Classical AI research is based on the capabilities of a program-controlled computer, which, according to Church's thesis, is in principle equivalent to a Turing machine. According to Moore's law, gigantic computing and storage capacities have been achieved, which only made possible the AI services of, e.g., the supercomputer WATSON (see Sect. 5.2). But, the power of supercomputers has a price that the energy of a small town can match. All the more impressive are the human brains that realize the power of WATSON (e.g., speak and understand a natural language) with the energy consumption of an incandescent lamp. By then, at the latest, one is impressed by the efficiency of neuromorphic systems that have evolved in evolution. Is there a common principle underlying these evolutionary systems that we can use in AI?

Biomolecules, cells, organs, organisms and populations are highly complex dynamic systems in which many elements interact. Complexity research is dealing interdisciplinary in physics, chemistry, biology, and ecology with the question how, by the interactions of many elements of a complex dynamic system (e.g., atoms in materials, biomolecules in cells, cells in organisms, organisms in populations), orders and structures can arise, but also chaos and decay.

© Springer-Verlag GmbH Germany, part of Springer Nature 2020
K. Mainzer, *Artificial intelligence – When do machines take over?*, Technik im Fokus,
https://doi.org/10.1007/978-3-662-59717-0_10

Generally, in dynamic systems, the temporal change of their states is described by equations. The state of motion of a single celestial body can still be precisely calculated and predicted according to the laws of classical physics. For millions and billions of molecules, on which the state of a cell depends, it is necessary to resort to high-performance computers that provide approximations in simulation models. It is remarkable that complex dynamic systems obey the same or similar mathematical laws in physics, chemistry, biology, and ecology.

The basic idea of complex dynamic systems is always the same [1, 2]: Only the complex interactions of many elements generate new properties of the overall system that cannot be traced back to individual elements. Thus, a single water molecule is not "moist", but a liquid due to the interactions of many such elements. Individual molecules do not "live", but a cell does because of their molecular interactions. In systems biology, the complex chemical reactions of many individual molecules enable the metabolic functions and regulatory tasks of entire protein systems and cells in the human body. In complex dynamic systems, we therefore distinguish between the micro-level of the individual elements and the macro-level of their system properties. This emergence or self-organization of new system properties becomes computable in systems biology and can be simulated in computer models. In this sense, systems biology is a key to the complexity of life.

In general, we imagine a spatial system consisting of identical elements ("cells") that can interact with each other in different ways (e.g., physically, chemically, or biologically) (Fig. 10.1 [3]). Such a system is called complex if it can generate non-homogeneous ("complex") patterns and structures from homogeneous initial conditions. This pattern and structure formation is triggered by local activity of its elements. This applies not only to, e.g., stem cells during the growth of an embryo, but also to transistors in electronic networks.

▶ We call a transistor locally active, when it can amplify a small signal input from the energy source of a battery to a larger signal output to generate non-homogeneous ("complex") voltage patterns in switching networks.

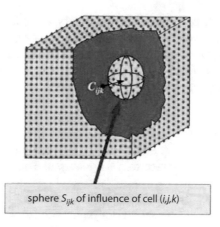

sphere S_{ijk} of influence of cell (i,j,k)

Fig. 10.1 Complex cellular system with locally active cells and local sphere of influence [3]

No radios, televisions, or computers would be without the local activity of such units. Important researchers such as the Nobel Prize winners I. Prigogine (chemistry) and E. Schrödinger (physics) were still of the opinion that a nonlinear system and an energy source are sufficient for structure and pattern formation. However, the example of transistors already shows that batteries and non-linear switching elements alone cannot generate complex patterns if the elements are not locally active in the sense of the described amplifier function.

The principle of local activity is of fundamental importance for the pattern formation of complex systems and has not yet been widely recognized. It can be defined mathematically in general, without having to rely on special examples from physics, chemistry, biology, or technology. We refer to nonlinear differential equations as known from reaction-diffusion processes (but by no means limited to liquid media as in chemical diffusion). We can clearly imagine a spatial lattice whose lattice points are occupied by cells that interact locally (Fig. 10.1). Every cell (e.g., protein in a cell, neuron in the brain, transistor in the computer) is mathematically a dynamic system with input and output. A cell state develops locally according to dynamic laws depending on the distribution of neighboring cell states. In summary, the dynamic laws are defined by the equations of the state of isolated cells and their coupling laws. In addition, initial and auxiliary conditions must be taken into account in the dynamics.

▶ In general, a cell is called locally active if a small local input exists at a cellular equilibrium point that can be amplified with an external power source to a large output. The existence of an input that triggers local activity can be systematically tested mathematically by certain test criteria. A cell is called locally passive if there is no equilibrium point with local activity. What is fundamentally new about this approach is the proof that systems without locally active elements do not, in principle, have complex structures and be able to create patterns.

Structure formation in nature and technology can be systematically classified by modelling application areas by reaction-diffusion equations according to the pattern described above. For example, the corresponding differential equations for pattern formation in chemistry (e.g., pattern formation in homogeneous chemical media), in morphogenesis (e.g., pattern formation of mussel shells, fur and feathers in zoology), in brain research (circuit patterns in the brain) and in electronic network technology (e.g., circuit patterns in computers) can be investigated.

In statistical thermodynamics, the behavior is determined by interaction of many elements (e.g., molecules) in a complex system. L. Boltzmann's 2nd law of thermodynamics only states that all structures, patterns, and orders decay in an isolated system if one leaves them to oneself. Thus, all molecular arrangements dissolve in a gas and heat is distributed uniformly and homogeneously in a closed space during dissipation. Organisms disintegrate and die if they are not in mass and energy exchange with their environment. But, how can order, structure, and patterns be created?

▶ The principle of local activity explains how order and structure are created in an open system through dissipative interaction or mass and energy exchange with the system environment. It complements the 2nd law as the 3rd law of thermodynamics.

Structural formations correspond mathematically to non-homogeneous solutions of the considered differential equations, which depend on different control parameters (e.g., chemical concentrations, ATP energy in cells, neurochemical messengers of neurons). For the considered examples of differential equations, we could systematically define the parameter spaces whose points represent all possible control parameter values of the respective system. In these parameter spaces, the regions of local activity and local passivity can be precisely determined, which either enable structure formation or are mathematically "dead". In principle, computer simulations can be used to generate the possible structure and pattern formations for each point in the parameter space (Fig. 10.2). In this mathematical model framework, structure and pattern formation can be completely determined and predicted.

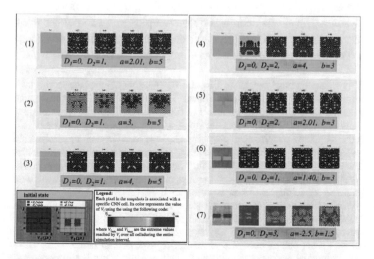

Fig. 10.2 Structure and pattern formation of a nonlinear diffusion and reaction equation [3]

Background Information
A completely new application of local activity is the "edge of chaos", where most complex structures arise. Originally stable ("dead") and isolated cells can be "brought to life" by dissipative coupling and trigger pattern and structure formation. Clearly spoken, they "rest" isolated at the edge of a stability zone until they become active through dissipative coupling. One could imagine isolated chemical substances resting in the hostile dark deep sea at the edge of a hot volcanic vent. The dissipative interaction of the originally "dead" elements leads to the formation of new forms of life. As chemical substances, however, they must carry within them the potential of local activity triggered by dissipative coupling.

This is unusual in that it seems to contradict the intuitive understanding of "diffusion": According to it, "dissipation" means that, e.g., a gas is distributed evenly and homogeneously in a space. However, not only unstable but also stable elements can trigger complex (inhomogeneous) structure and pattern formation by dissipative coupling. This can be proven exactly for nonlinear reaction and diffusion equations. In the parameter spaces of these equations, the "edge of chaos" can be marked as part of the region of local activity.

Even the human brain is an example of a complex dynamic system in which billions of neurons interact neurochemically. Complex switching patterns are created by multiple electrical impulses, which are associated with cognitive states such as thinking, feeling, perceiving, or acting. The emergence of these mental states is again a typical example of self-organization of a complex system: The single neuron is quasi "stupid" and can neither think nor feel nor perceive. Only their collective interactions and interconnections under suitable conditions generate cognitive states.

The neurochemical dynamics between the neurons take place in the neuronal networks of brains. Chemical messengers cause neuronal state changes through direct and indirect transmission mechanisms of great plasticity. Different network states are stored in the synaptic connections of cellular switching patterns (cell assemblies). As is usual in a complex dynamic system, we also differentiate in the brain between the micro states of the elements (i.e. the digital states of "firing" and "non-fire" when a neuron is discharged and at rest) and the macro states of pattern formation (i.e. switching patterns of jointly activated neurons in a neuronal network). Computer visualizations (e.g., PET images) show that different macroscopic circuit patterns are correlated with different mental and cognitive states such as perception, thinking, feeling, and consciousness. In this sense, cognitive and mental states can be described as emergent properties of neural brain activity: Individual neurons can neither see, feel, nor think, but brains connected to the sensors of the organism.

Current computer simulations therefore observe pattern formation in the brain, which we attribute to nonlinear system dynamics, the local activity of neurons, and the action potentials they
trigger. Their correlations with mental and cognitive states are
revealed on the basis of psychological observations and measurements: Whenever people see or speak this or that, pattern formation can be observed in the brain. In brain reading, individual
patterns can now be determined to such an extent that the corresponding visual and auditory perceptions can be decoded from
these circuit patterns using suitable algorithms. However, this
technique is only in its infancy.

▶   In a top-down strategy, neuropsychology and cogni-
     tive research are investigating mental and cognitive
     abilities such as perception, thinking, feeling, and
     consciousness, and try to connect them with corre-
     sponding brain areas and their interconnection pat-
     terns. In a bottom-up strategy, neurochemistry and
     brain research investigate the molecular and cellular
     processes of brain dynamics and explain neuronal
     brain interconnection patterns, which in turn are cor-
     related with mental and cognitive states [4].

Both methods suggest a comparison with the computer, in
which, in a bottom-up strategy, the meanings of higher user languages of humans is derived from the "machine language" of the
bit states in, e.g., transistors, while in a top-down strategy, conversely, the higher user languages are translated to the machine
language via various intermediate stages (e.g., compiler and
interpreter). However, while in computer science, the individual
technical and linguistic layers from the interconnection level via
machine language, compiler, interpreter etc. to the user level can
be precisely identified, brain and cognitive research has so far
only been a research program.

  In brain research, so far only the neurochemistry of neurons
and synapses and the pattern formation of their circuits are well
understood, i.e. the "machine language" of the brain. The bridge
(middleware) between cognition and "machine language" has

yet to be reconstructed. This will require many more detailed empirical studies. It is by no means already clear whether individual hierarchical levels can be precisely distinguished as in computer design. Apparently, the architecture of brain dynamics proves to be much more complex. In addition, the development of the brain was not based on a planned design, but on a multitude of evolutionary algorithms that were developed more or less randomly under different conditions over millions of years and are connected to each other in a complex way.

In complexity research, the synaptic interaction of neurons in the brain can be described by coupled differential equations. The Hodgkin-Huxley equations are an example of nonlinear reaction diffusion equations that can be used to model the transmission of nerve impulses. These equations have been struck down by the medicine-Nobel Prize winners A. L. Hodgkin and A. F. Huxley by empirical measurements and provide an empirically confirmed mathematical model of neuronal brain dynamics.

Example

In Fig. 10.3, the information channel (axon) of a neuron (a) is represented by a chain of identical Hodgkin-Huxley (HH) cells coupled by diffusion compounds (b). These couplings are technically represented by passive resistors. The HH cells correspond to an electrotechnical interconnection model (c): In a biological nerve cell, ionic currents of potassium and sodium alter the voltages on the cell membrane. In the electrotechnical model, sodium and potassium ion currents are triggered together with a current outflow by an external axon membrane current. The ion channels are technically realized by transistor-like amplifiers. They are connected to a sodium ion and potassium ion battery voltage, a membrane capacitor voltage and a voltage discharge. In this way, the input flows can be strengthened according to the principle of local activity, in order to create a potential for action if a threshold value is exceeded ("fire"). These action potentials trigger chain reactions that lead to interconnection patterns of neurons.

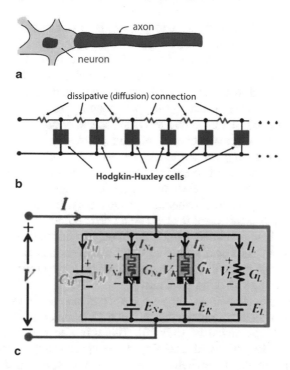

Fig. 10.3 **a** axon of a neuron, **b** electrotechnical model of an axon, **c** electrotechnical model of the Hodgkin-Huxley equations. I external axon membrane current, E membrane capacitor voltage, I_{N_a} sodium ion current, E_{N_a} sodium ions battery voltage, I_K potassium ion current, E_K potassium ions battery, I_L leakage current, E_L leakage battery voltage [3]

As already explained, such differential equations can be used to precisely determine the corresponding parameter spaces of a dynamic system with locally active and locally passive regions. In the case of the Hodgkin-Huxley equations, we obtain the parameter space of the brain with the precisely measured regions of local activity and local passivity. Action potentials of neurons that trigger circuit patterns in the brain can only develop in the area of local activity. Computer simulations can be used to systematically investigate and predict these circuit patterns for the various parameter points.

Thus the region at the "edge of chaos" can be determined exactly. It is tiny and is less than 1 mV and 2 μA. This region is associated with large local activity and pattern formation, which can be visualized in the corresponding parameter spaces. Therefore, it is assumed that this is an "island of creativity".

For an electrotechnical realization, however, the original equations of Hodgkin and Huxley proved to be faulty. The physicians Hodgkin and Huxley interpreted some switching elements in a way that led to electrotechnical anomalies. For example, they assumed a time-dependent conductivity (conductance) to explain the behavior of the potassium and sodium ion channels. In fact, these temporal changes could only be calculated numerically from empirically derived equations. Theoretically, it was not possible to explicitly define corresponding time functions for time-varying switching elements.

The anomalies dissolve when the ion channels are explained by a new switching element that Leon Chua had already mathematically predicted in 1971 [5]. This refers to the memristor (from the English word "memory" for memory and "resistor" for resistor). With this switching element, the electrical resistance is not constant, but depends on its past. The actual resistance of the memristor depends on how much charge has flowed in which direction. The resistance is maintained even without energy supply. This realization has enormous practical consequences, but could also be a breakthrough for neuromorphic computers oriented to the human brain. First, we explain the concept of a memristor.

In practice, computers equipped with memristors would be ready for operation immediately after switching on without booting. A memristor retains its memory content if it is read with alternating current. A computer could therefore be switched on and off like a light switch without information being lost.

Background Information

Traditionally, in electrical engineering, only resistance, capacitor, and coil were distinguished as switching elements. They combine the four switching variables charge, current, voltage, and magnetic flux: Resistors connect charge and current, coils connect magnetic flux and current, capacitors connect voltage and charge. But, what connects charge and magnetic flux? L. Chua postulated the memristor in 1971. Mathematically, this is done with a function R (q) ("memristance function") is defined, in which the change of the magnetic flux Φ with the charge q is being held, i.e.

$$R\left(q\right) = \frac{d\Phi\left(q\right)}{dq}.$$

The temporal change of the charge q defines the current $i(t)$ with

$$i\left(t\right) = \frac{dq}{dt}.$$

The temporal change of the magnetic flux Φ defines the voltage $v\left(t\right)$ with

$$v\left(t\right) = \frac{d\Phi}{dt}.$$

The result is that the voltage v at a memristor about the current i depends directly on the memristance:

$$v = R\left(q\right) i.$$

This is reminiscent of Ohm's law $v = R\,i$ which defines that the voltage v is proportional to current i with the resistance R as a proportionality constant. However, the memristance is not constant, but depends on the state of the charge q. Conversely, for electricity, it holds

$$i = G\left(q\right) v,$$

where the function $G\left(q\right) = R(q)^{-1}$ is called "memductance" (composed of the English word "memory" for memory and "conductance" for conductivity).

A memristor can be generalized as a memristive system. A memristive system is no longer reduced to a single state variable and a linear charge- or flow-driven equation.

▶ **Definition** A memristive system is an arbitrary physical system, which is defined by a set of internal state variables $\vec{s}$ (as a vector). It follows a general input-output equation

$$\vec{y}\,(t) = g\,(\vec{s},\,\vec{u},\,t)\,\vec{u}\,(t)$$

with the input $\vec{u}\,(t)$ (z. B. Voltage) and the output $\vec{y}\,(t)$ (z. B. Electricity). The development of a state is generally determined by a differential equation:

$$\frac{d\vec{s}}{dt} = f(\vec{s},\,\vec{u},\,t)$$

Memristive systems exhibit exceptionally complex and nonlinear behavior. Typical is the hysteresis curve in the v/i-diagram in Fig. 10.4. It runs in closed loops through the pinched hysteresis loop [6].

▶ In general, hysteresis refers to the behavior of the output variable of a system that reacts to an input variable with a delayed (Greek hysteros) signal and varies. The behavior does not only depend directly on the input variable, but also on the previous state of the output variable. If the input variable is the same, the system can adopt one of several possible states.

As "neuristors", memristive systems simulate the behavior of synapses and therefore become interesting for neuromorphic computers. For this purpose, the ion channels in the circuit model in Fig. 10.3 are regarded as memristive systems. Hodgkin-Huxley's time-dependent conductivity G_k of the potassium ion channel is replaced by a charge-controlled memristor

Fig. 10.4 Hysteresis curve of a memristor (depending on the angular frequency) ω with $\omega_1 < \omega_2$

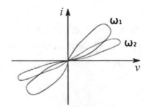

dependent on a state variable. Hodgkin-Huxley's time-dependent conductivity G_{Na} of the sodium ion channel is replaced by a charge-controlled memristor which depends on two state variables. These well defined quantities explain precisely the empirical measurement and observation data of synapses and neurons [7].

But how can such neuristors be technically realized? R. Stanley Williams of the company Hewlett-Packard (Silicon Valley) has constructed a version for the first time in 2007, which has meanwhile been constantly simplified and improved [8]. Imagine a crossbar network of intersecting vertical and horizontal wires, reminiscent of a wire mesh (Fig. 10.5 [9]). The intersections of a vertical and horizontal wire are connected by a switch. To close the switch, a positive voltage is applied to both wires. To open it, the charge is reversed.

Background Information

In order to achieve memrestive behavior, the switches are constructed according to a specific architecture. It is reminiscent of a sandwich in which a titanium dioxide layer a few nanometers thick lies between two platinum electrodes (as "slices of bread"). In Fig. 10.5, the lower titanium dioxide layer serves as an insulator. The upper titanium dioxide layer has oxygen deficiencies. You can imagine them as small bubbles in a beer—with the difference that they cannot escape. This titanium oxide layer has a high conductivity. If a positive voltage is applied, the oxygen deficiencies shift. This reduces the thickness of the lower insulation layer and increases the overall conductivity of the switch. A negative charge, on the other hand, attracts the positively charged oxygen deficiencies. This increases the insulation layer and reduces the overall conductivity of the switch.

The memristive behavior becomes apparent when the voltage is switched positively or negatively: Then the small bubbles of the oxygen deficiencies do not change, but remain where they are. The border between the two titanium dioxide layers is "frozen", so the switch can "remember" how much voltage was last applied. It works like a memristor.

Other memristors use silicon dioxide layers a few nanometers in size, which require only low costs. The crossbar memories manufactured by Hewlett-Packard already have an enormous packing density of approx. 100 Gibit/cm^2. They could also be combined with other semiconductor structures. Therefore, it cannot be excluded that they initiate the development of neuromorphic structures to simulate the human brain.

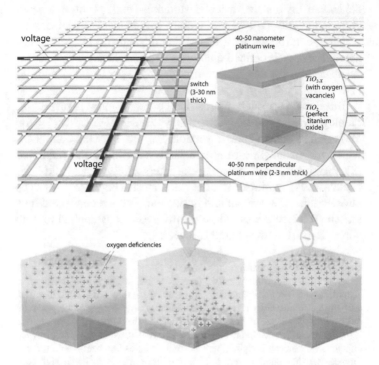

Fig. 10.5 Memristive system with switches made of titanium dioxide [9]

The outcome of this research program was the mathematical Hodgkin-Huxley model of the brain. In the Human Brain Project of the EU, an exact empirical modelling of the human brain with all neurological details is aimed at. With the technical development of neuromorphic networks, an empirical test bed would be available for this mathematical model, in which predictions about pattern formation in the brain and their cognitive meanings can be verified.

From psychology, we know that mental and cognitive states interact in an extremely complex way. Perceptions can thus trigger thoughts and ideas that lead to actions and movements. However, a perception is usually also connected with a self-perception: It's me who perceive. Self-awareness, combined with the storage of one's own biography in memory, leads to ego-consciousness. If all these different mental states are associated with circuit patterns in the brain, then not only the interactions of individual neurons must be recorded, but those of cell assemblies with cell assemblies of cell assemblies etc.

In principle, differential equations can also be introduced, which do not depend on the local activities of individual neurons, but on whole cell assemblies, which in turn can depend on cell assemblies of cell assemblies etc. This results in a system of nonlinear differential equations which are interconnected on different levels and thus model an extremely complex dynamic. Connected to the sensors and actuators of our organism, they record the processes that create our complex motor, cognitive, and mental states. As already stressed, we do not yet know all these processes in detail. But, it is clear how, in principle, they can be mathematically modelled and empirically tested in neuromorphic computers.

10.2 Natural and Artificial Intelligence

In evolution, networks first emerged as subcellular supply, control, and information systems in complex gene and protein networks. Nerve cells eventually led to the development of cellular information, control, and supply systems based on neurochemical signal processing. Examples are ant populations, human brains, and cyberphysical systems of human society.

▶ According to the working definition in chap. 1, a system is called intelligent if it can solve problems independently and efficiently. Traditionally, a distinction is made between natural systems, which were created in evolution, and technical ("artificial") systems, which were introduced in the history of technology. The respective degree of intelligence depends on the complexity of the problems that can be measured in mathematical complexity theory.

In evolution, effective problem-solving methods developed without symbolic representation in computer models. Subcellular, cellular, and neuronal self-organization created the corresponding complex networks. In principle, they can be simulated by computer models. These simulations are based on a fundamental mathematical equivalence of neural networks, automata, and machines.

It can be proven that a McCulloch-Pitts network with integers as synaptic weights (see Sect. 7.2) can be simulated by a finite automaton (see Sect. 5.2). Conversely, the performance of a finite machine can also be achieved by a McCulloch-Pitts network. In other words, an organism equipped with a neuronal nervous system of the McCulloch-Pitts network type can only deal with problems of complexity that a finite automaton can handle. In this sense, such an organism would be as intelligent as a finite automaton.

But, which neuronal networks correspond to Turing machines which, according to Church's thesis, are regarded as prototypes of program-controlled computers?

▶   It can be proven that Turing machines simulate neuronal networks whose synaptic weights are rational numbers ("fractions") and have feedback loops ("recurrent"). Conversely, Turing machines can be precisely simulated by recurrent neuronal networks with rational synaptic weights [10].

In the biological model, the numerical values of the weights correspond to the chemical strengths of synaptic compounds that are altered by learning algorithms of neuronal networks. Intense synaptic couplings produce neural interconnection patterns that correspond to mental, emotiona, or motor states of an organism. If we take a Turing machine as a prototype of a program-controlled computer, then, according to this proof, a brain with finite synaptic strengths can be simulated by a computer. Conversely, the processes in a Turing machine (i.e. a computer) can be reproduced by a brain with finite synaptic intensity variables. In other words, the intelligence level of such brains corresponds to the intelligence level of a Turing machine.

In practice, it follows that neural networks of this kind can, in principle, be simulated on a suitable computer. In fact, neural networks for practical applications (e.g., pattern recognition) are still largely simulated on computers. Only neuromorphic computers would replicate neural networks directly.

But what can neural networks with synaptic weights do, which do not only allow rational numbers (i.e. finite quantities such as 2,3715 with a finite number of decimal fractions), but also any real numbers (i.e. decimal fractions with an infinite number of digits after the comma such as 2,3715 ..., which are also not computable)? Technically speaking, such networks would not only perform digital but analog calculations.

▶ **Definition**

In signal theory, an analog signal is a signal with an infinitely variable and uninterrupted course. Mathematically, an analog signal is defined as a smooth function that can be infinitely differentiated, i.e. in particular, it is continuous. The graph of such a function has no corners and interruptions that cannot be differentiated. This allows the time continuous course of a physical quantity to be described in the form of an analog signal.

An analog-to-digital converter discretizes a time-continuous input signal into individual discrete samples.

In fact, many processes in a natural organism can be understood as analogous. For example, signal processing during vision is described by continuous electromagnetic fields that hit sensors. The acoustics of hearing are also based on constant waves. Even under pressure, the skin sensors convey a continuous and not a digital sensation. Now it will be objected that measured values in a finite physical world are finite and therefore in principle digitizable.

The theoretical consequences of analog neural networks have fundamental importance for artificial intelligence. Mathematically, analogue neuronal networks can be defined uniquely with any real numbers as synaptic weights, if the mathematical theory of real numbers is assumed. The central question is whether analog neural networks can do "more" than neural networks with rational numbers and thus "more" than Turing machines or digital computers. This would be a central argument in the AI debate, according to which mathematics is "more" than computer science and cannot be reduced to digital computers.

A central achievement of automata and machines is the recognition and understanding of formal languages (cf. chap. 5). An automaton recognizes a read word as a formal sequence of symbols, if it changes after finite many steps into an accepting state and stops. A language accepted by a machine consists only of words that can be recognized by the machine. By that, it can be proven that finite automata recognize exactly the regular languages (see Sect. 5.2). Context-free languages use rules whose derivation of words does not depend on surrounding symbols. They are recognized by more powerful pushdown automata. After all, recursively enumerated languages are so complex that they can only be recognized by Turing machines.

Thus, neural networks with rational synaptic weights (like Turing machines) can also recognize recursively enumerated languages. These can be natural neuronal systems of organisms as well as artificial neuromorphic computers which correspond to the laws of recurrent neural networks with rational synaptic weights. It can now be proven:

▶   In principle, analog neural networks (with real synaptic weights) can also recognize non-computable languages in exponential time [11].

Such proofs are mathematically possible if the concept of computability is extended from natural (and rational) numbers to real numbers [12]. Instead of digital processes with difference equations, it is also possible to describe continuous real processes with differential equations. In other words: All kinds of dynamic systems like, e.g., flows in physics, reactions in chemistry, and organisms in biology can be represented by corresponding extended analog systems with real numbers.

Background Information

However, analog neural networks are not expected to solve NP-hard problems in polynomial time (see Sect. 3.4). Thus it can be proved that, for example, the problem of the salesman is also NP on the real numbers NP.

On the other hand, according to a proof of the logician A. Tarski (1951), any set definable on real numbers is also decidable. On the other hand, there are quantities that can be defined on the whole numbers, which cannot be decided.

This is a consequence of Gödel's incompleteness theorem of arithmetic (see Sect. 3.4). The real computability is apparently partly "simpler" than the digital computability on the whole number.

The advantage of computability generalized to real numbers (analog computability) is in any case that it captures analogue processes in organisms, brains, and neuromorphic computers more realistically. Thus, a fundamental equivalence of evolutionary, mathematical, and technical procedures becomes obvious, which suggests an extension of Church's thesis:

▶ **Important**

Not only digital effective methods can be represented by computer models in the sense of a (universal) Turing machine, but also analog effective methods of nature. If this extended thesis of Church is correct, then the invention of the computer opens up a fundamental insight, the scope of which was not foreseeable at first:

All effective dynamic processes (both natural and technical or "artificial") can be modelled on a (universal digital or analog) computer.

That would be the core of a unified theory of complex dynamic systems. The symbolic codes with numbers in the computer would only be our way of information processing representing atomic, molecular, cellular, and evolutionary processes.

Degrees of computability can be distinguished: A non-deterministic Turing machine, for example, uses random decisions in a computation in addition to the usual effectively computable elementary operations. For this, we extend the concept of the Turing machine by the (Turing based) concept of the Ψ-oracle machine [13–15]:

▶ **Definition**
For a Ψ-oracle machine, in addition to the commands of a (deterministic) Turing machine, an operation Ψ (e.g., "replace the numeric value x by $\Psi(x)$") of which we do not know whether it is computable. The computation depends on the "oracle" Ψ.

An example in nature would be a mutation as a random change in the effective processing of DNA information. This extension computability is also called relative computability:

A function is computable relative to Ψ if it is computable by a Ψ-oracle machine.

Accordingly, a relativized version of Church's thesis can be formulated: All relative to Ψ effective processes can be simulated by a (universal) Ψ-oracle Turing machine.

Accordingly, an extended analogous version (for real numbers) of Church's thesis can also be formulated.

It can be proven that, in polynomial time, an analog neural network recognizes the same class of languages that a suitable Ψ-oracle Turing machine recognizes in polynomial time.

Acording to our definition of artificial intelligence in chap. 1, it follows that a natural organism with a corresponding analog neuronal nervous system or a corresponding technical neuromorphic system is as intelligent as this Ψ-oracle Turing machine.

Some mathematical and natural objects like, e.g., a sequence of zeros or a perfect crystal are intuitively simple, other objects like the human organism or the sequence of numbers of a random decimal fractions like e.g. 0,523976 … apparently have a complex evolutionary history. The complexity of these objects can be made more precise by their logical depth, i.e. the computational time a universal Turing machine can use to generate its development process from an algorithmically random input. Computational time is not a physical measure of time, but a logical-mathematical measure of complexity, which determines the number of elementary arithmetic operations of a Turing machine depending on the input.

With natural objects, the algorithmically random input corresponds to the more or less random initial data of evolution. This definition of complexity through logical depth of the development process is thus independent of the respective technical standard of a computing machine. It can be shown [16] that (complex) objects with logical depth cannot be created "quickly" from simple objects—neither with a deterministic nor with a probabilistic process. This proof theoretically confirms our empirical knowledge of the evolution of life, whose complex organisms have developed through many complex and more or less random phase transitions (bifurcations).

The transfer of logical depth to physical and evolutionary complexity of life is based on the assumption of Church's extended thesis, according to which developmental processes in nature can be simulated with appropriate efficiency by computer models and thus (extended) Turing machines.

Natural processes are often modelled by continuous differential equations. Digital machines cannot solve continuous differential equations of dynamic systems exactly (Occasionally also the concept of computability is not sufficiently robust for continuous system laws, since a computable differentiable function can have a non-computable derivation). But, digital computing methods can approximate dynamic processes with finite precision. Even for stochastic phase transitions, as they typically occur in complex dynamic systems and are mathematically described by stochastic differential equations (e.g., Master equations), discrete stochastic models are known that can be simulated on computers.

Since computer programs are invented by humans and must be understandable for humans, they are represented with symbols of programming languages. However, this is only a special coding of information in technical systems. In biological intelligent systems, such an intermediate representation with speech symbols is not necessary, since information is encoded and understood by molecular and cellular interactions. The neurochemical signal exchange of organisms and neurons is organized according to the nonlinear laws of complex dynamic systems.

The intelligent performances of the entire system are not recognizable from the individual signals of the organisms in populations or the neurons in brains. Thus, even the electrical impulses and voltage states of a computer do not allow to deduce their processing of information and knowledge. This requires translation programs over several layers from information and knowledge representation to machine language, which corresponds to the technical-physical signals.

In humans, knowledge is additionally connected with consciousness. Corresponding data and rules are loaded from the long-term memory into the short-term memory and can be symbolically represented there: Then I know it's me who knows, can, or does something. In principle, it cannot be ruled out that AI systems will be equipped with consciousness-like capabilities in the future. Such systems would create their own experiences and identities, which are quite different from our human self-experiences. People also develop different mentalities in different social contexts, which distinguish them individually, although they have the same information system of the brain. Therefore, the technical modulation of consciousness-like functions only makes limited sense if it is required for service tasks of AI systems.

However, a sole fixation of AI research on an AI system with human-like consciousness would be a dead end. Intelligence only emerges in interaction with a corresponding environment. Physiologically, the human brain has hardly changed since the Stone Age. We only become people of the twenty-first century through our possibilities of interaction in this technical society.

The globalized knowledge society itself becomes a complex intelligent system, in which a variety of more or less intelligent functions are integrated and the individual human being with his or her awareness of the world and his or her environment is an acting part. Therefore, cyberphysical systems aim at the implementation of social and situational knowledge in AI systems, in order to improve their service tasks in this world in dealing with people. The motto is therefore: Cyberphysical systems with distributed artificial intelligence instead of isolated artificial intelligence of single highly trimmed robots or computers!

In practice, social and situational knowledge is comprehensible by declarative knowledge representation only in a limited way, although in principle possible according to Church's extended thesis. In this case, we make it simple like nature: we orient ourselves on implicit knowledge that is applied without rule-based representation. We simply use the devices in a telematically networked world of Cyberphysical Systems without being aware of their programming. With their self-explanatory user interfaces, they are embedded in this technical world just as ergonomically as the plough in the field or the hammer and anvil in a forge in earlier life worlds.

Therefore, a one-sided research program of the AI can be criticized, which is limited to the construction of a robot with human-like consciousness and as complete as possible knowledge representation of its outside world. We should rather equip our human world with ubiquitously distributed AI functions in cyberphysical systems and thus make it more worth living.

For robotics, R. A. Brooks from the AI Laboratory of the MIT had propagated "intelligence without representation", which developed during evolution as swarm intelligence for insect populations. Instead of robots, which have to decode motoric knowledge representation at a high level of programming languages, simple machines are used, whose processors interact without rigid program sequences. Intelligent problem solutions of these robot populations are collective achievements without knowledge representations of the individual machines [17].

In the future, an integrated strategy combining hybrid AI systems with knowledge-based programming and situational learning will certainly have to be pursued. Only in this way, it will be possible for robots not only to skillfully coordinate their actions with each other, but also to plan and decide how more sophisticated biological systems can be. In cognitive and AI research, there is a growing realization that the role of consciousness in human problem solving has been overestimated and the role of situational and implicit learning underestimated. According to this, intelligence is an interactively developing ability and not a static and rigidly programmed property of an isolated system.

Societies are extremely complex systems of interacting individuals, institutions, and subsystems whose nonlinear dynamics develop in phase shifts. Because of the many variables on which their development depends, they are also referred to as high-dimensional systems. Like biological organisms with their billions of interacting cells, organs, and nervous systems, social systems can be understood as superorganisms equipped with economic metabolic cycles and extrasomatic information systems. This metaphor of sociobiology can be made more precise with the theory of complex dynamic systems

Examples are dynamic economic models of economic systems, modelling of transport networks, energy supply systems or dynamic models of the Internet. Companies therefore grow together with cyberphysical systems in complex dynamic communication systems. Complex dynamic systems organize themselves under suitable marginal and initial conditions. It is important to influence the appropriate control values so that the complex system develops itself in the desired way.

The dynamics of human societies are far more complex than gene and protein networks, cellular organisms, brains, and animal populations, because they are the interactions of conscious people with their own will. So, people are not only captured and driven like molecules in a fluid stream by collective trends and vortices. In unstable situations, as history shows, few people can change global dynamics, be it through political revolutions or technical-scientific innovations. Every day, millions of people are willingly or unwillingly involved in creating trends in the global social and economic dynamics of a society. In the process, various feedback mechanisms take place between people and their social environment, which in turn can trigger unconscious side effects. Thus, distributed artificial intelligence creates an extremely complex communication and supply system whose dynamics are determined by technical, economic, social and cultural networks.

▶ The laws of complex dynamic systems, which can be modelled in computer models, emerge as the basis of a unified theory of complex networks. [18, 19]. Systems biology and evolutionary biology, brain and cognitive research as well as software and hardware development of computers, robots and other devices, cyberphysical systems, and telematically networked societies are the first steps in this direction of global development of our Earth system.

Because of Gödel's incompleteness theorem, there will not be a supercomputer that could formally represent all knowledge in its entirety. Incomplete systems, however, can be extended step by step in order to open up ever richer representations of knowledge without limitation. The laws of complex dynamic systems allow us to estimate trends, critical phases, and attractors of development. The scientific challenge of systems research is then to better understand the complex network dynamics of cyberphysical systems and communication systems of human society.

In practice, as H. Simon, as one of the founding fathers of the AI, showed with the example of economy, intelligent systems act under the conditions of bounded rationality. This context- and situation-dependent limitation is not a fundamental limitation of knowledge. In the sense of Gödel's incompleteness theorem, they can, in principle, be overcome in order to reach new revisable limits.

Knowledge is created by constructing models of the outside world, which are generated by methods and programs, organizations and institutions of complex dynamic systems. Even biological systems such as humans have knowledge as mental constructions in their heads and not as mirror images of the outside world. Collective information systems such as human societies generate their collective knowledge as a social construction. Limited representations of knowledge are thus system constructions. With this self-created world in mind, intelligent systems operate in open information spaces under the dynamically changing conditions of bounded rationality.

10.3   Quantum Computer and Artificial Intelligence

So far we looked at artificial intelligence on machines of classical physics. With quantum computing we go back to the smallest units of matter and the limits of natural constants like Planck's quantum of action and the speed of light—the ultima ratio of a computer. As a physical machine, the performance of a computer depends on the circuit technology used. Their growing miniaturization has delivered new generations of computers with increasing memory capacity and reduced computing time. However, increasing miniaturization leads us into the order of magnitude of atoms, elementary particles, and smallest energy packets (quanta), to which our usual laws of classical physics apply only to a limited extent. Instead of classical machines according to the laws of classical physics, quantum computers would then have to be used which function according to the laws of quantum mechanics [20].

Quantum computer would lead to breakthroughs in information and communication technology with an enormous increase in computing capacity. Problems like, e.g., the factorization problem, which until now had exponential complexity and thus were practically unsolvable, will then be solvable polynomially. Technically, quantum computers would therefore lead to an immense increase in our problem-solving capacities. In the sense of the complexity theory of computer science, the high computing times of individual problems could be considerably shortened (e.g., with polynomial computing time, although they do not belong to complexity class P in classical computers). But, could quantum computers also realize non-algorithmic thought processes beyond the complexity limit of a universal Turing machine? Would they open up new possibilities for artificial intelligence?

Let us first recall some basic properties of quantum physics [21]:

Background Information

A quantum object (e.g., photon) behaves like a particle as long as we measure a particle property such as location or momentum. If one measures a wave property such as the frequency of light, the quantum object (e.g., photon) behaves like a wave. Therefore, whether the quantum object is a wave or a particle is not fixed from the outset in classical physics, as is the case with balls or water waves, but is decided only by the respective experimental measuring arrangement. This wave-particle dualism of quantum physics differs fundamentally from the world of classical physics.

In Bohr's atomic model, the superposition of two states of an electron clearly means that the electron is on two different orbits at the same time. This uncertainty lasts until the electron emits or absorbs a photon after a certain time and thus commits itself to one of its states. This happens during an interaction, e.g., with a laser pulse. Two waves that oscillate in a common mode like a single wave are also called coherent. The process by which they are brought into their own state is called decoherence.

▶ If we use hydrogen atoms to store information, then in addition to the ground state, energy is used to store E_0 for 0 and the excited state with energy E_1 for 1, an intermediate state shall also be considered in which the wave of the ground state and that of the excited state are superimposed with the same amplitude. Such a quantum bit (qubit) is half 0 and half 1, whereas a classic bit is always either 1 or 0.

Example

Erwin Schrödinger illustrated superimposed quantum states in a thought experiment with a cat, which is locked into a box together with a hydrogen cyanide bottle. A hammer mechanism is associated with a random process such as the decay of an atomic nucleus. If the atomic nucleus decays, the hammer mechanism is triggered, the hydrogen cyanide bottle is destroyed, and deadly poison is released. But, nobody can predict whether the atomic nucleus will decay and the cat will be dead or alive. According to Schrödinger, the cat in the box is in a superimposed state of dead and alive, corresponding to the superimposed states of the quantum states "decay" and "non-decay" of the atomic nucleus. Only by measurement and observation, i.e. by opening the box the overlay state is lifted and the cat is either "dead" or "alive". One then also speaks of the "collapse" or "reduction" of the superimposed "wave package" of both partial states "dead" and "alive". The wave packet of the superposition mathematically corresponds to a probability amplitude for both partial states.

For the technical construction of quantum computers, there are great possibilities, but also considerable problems of realization. Besides the tiny size of atomic switches, their enormous switching and signal speed and their low energy consumption, quantum computers could be used for the simultaneous (parallel) processing of large data masses. The reason is the superposition principle of quantum physics, which allows the formation of quantum bits. In serial data processing, a decision to use a large mass of data must be considered consecutively for each data unit.

Background Information
With parallel data processing, the decision algorithm can work for all data simultaneously. For a quantum computer, all possible input bits are set to a superposition state of 0 and 1 in equal proportions. After processing this input in the atomic circuit technology of the quantum computer, we get a superposition of all possible outputs of this calculation. This simultaneous processing of all possible inputs is called quantum parallelism. Some authors have compared quantum parallelism with the superimposition of sound waves from different musical instruments playing simultaneously in an orchestra. We never receive the melody of a single instrument that plays a sequence of notes in series, but only in superimposition with other sequences of notes.

However, there is a fundamental difference between the superposition of acoustic waves and quantum waves. Quantum waves are probability amplitudes. The intermediate state of superimposed quantum bits jumps randomly into binary bit states (i.e. 0 or 1) when interacting with the outside world (e.g., laser pulse of a reading device). In contrast to acoustic waves, a single "quantum melody" changes when we "read" it from the superposition of all quantum melodies. In quantum physics, we say that the coherence of the superposition states is lost by interaction with the outside world (e.g., observation and measurement processes) (decoherence). For quantum computers, this causes major technical problems, such as how quantum bits (as coherent quantum states) can be stably stored without changing uncontrollably and randomly due to external disturbances (e.g., interaction with materials).

The laws of quantum mechanics have practical consequences for the computing of computers. For example, if we have to solve two subtasks of a problem, then a classic computer has to solve one subtask after the other (serially) first and then the other. In a quantum computer, however, the two subtasks could be combined and simultaneously processed as a superposition of states. Analogous to parallel computers with several processors, we speak of quantum parallelism.

Example

As an example, let's consider a task where a computer has to find a natural number with a certain property. A classical computer enumerates the numbers 1, 2, 3, ... and checks successively whether the respective number has the required property. If the number n you are looking for is very large, then the criterion must be tested n times and thus enormous computing time can be consumed. A quantum computer could check the criterion for a large number of numbers simultaneously and thus only once.

As usual, decimal numbers are represented by binary numbers corresponding to bit sequences. In the quantum computer a bit is represented by an alternative quantum state of a quantum system. As an example, we choose the alternative spin of an elementary particle, which can be left or right spinning. 0 should correspond to one and 1 to the other spinning device. A bit sequence then represents a sequence of rotating elementary particles. Depending on their spins, a combination of binary states of, e.g., seven particles can represent 2^7 possibilities such as 0000000 (for the decimal number 0), 0000001 (for the decimal number 1), 0000010 (for the decimal number 2) etc., i.e. any number between 0 and 127.

In a classical computer, the dual numbers 0000000, 0000001, 0000010, ... would have to be entered one after the other and then checked for the required criterion. The spins can be transported by sufficiently strong energy impulses into the opposite spinning device. With weak energy impulses, however, the particle only changes its spin sometimes,

sometimes not. In this case, there is a coincidence and we can only make probability statements about the spin behavior.

While Schrödinger's cat was simultaneously dead and alive in a closed box, the particle is then correspondingly, as long as it is unobserved and unmeasured, in a superimposed state (superposition) of opposing spins. If all seven particles are fired with weak energy impulses, then all seven particles are in superimposed states as long as they are not observed and measured. In this superposition, they can represent all 128 different states and thus numbers at the same time.

So, if a quantum computer with these seven particles is prepared in this superposition, it can check the required criterion at once for all 128 numbers simultaneously. It is easy to realize that even a few hundred particles can represent gigantic numbers at the same time and thus lead to unimaginable computing speeds today.

However, if individual values are read out, the superposition (probability amplitude) collapses "randomly" into its partial states and is reduced to the special values of its partial states.

A quantum computer works according to the laws of quantum physics [22], according to which the output of quantum states is clearly computable on the basis of the entered quantum states, as long as their coherence is not disturbed. In quantum physics, a quantum state develops in time clearly determined according to the Schrödinger equation, which is a deterministic differential equation. So far the computational process of a quantum computer can be understood according to the model of a deterministic Turing machine just like other generations of computers on a mechanical, electromechanical or electronic basis [23]. Due to quantum parallelism, however, gigantic amounts of data can be processed simultaneously by a quantum computer at lightning speed, which can be transformed into the superposition of a single quantum state. When reading the individual data, a random process occurs that cannot be predicted, in principle. This makes quantum computers a non-deterministic Turing machine.

In the previous Sect. 10.2, a hierarchy of machines and automata was presented, which correspond to neural networks of increasing efficiency. Turing machines are mathematically equivalent to neural networks with rational numbers as synaptic weights. They can recognize recursive languages determined by Chomsky grammars. Analog networks with real numbers as synaptic weights correspond to special oracle machines, i.e. Turing machines, which are extended by (polynomially limited) oracles and can even recognize non-recursive languages.

▶ Quantum computers are non-deterministic oracle machines that rely on quantum oracles. Quantum oracle means the random reduction of the wave packet (superposition of data), which occurs when the data of the machine output is read out.

Quantum computers can also be characterized by cellular quantum automata or neural quantum networks [24, 25]. This raises the question of which model of neuronal networks can be used to describe the human brain.

Background Information

Roger Penrose, a British mathematician with significant contributions to mathematical physics and cosmology, tried to understand the human brain as a special kind of quantum computer [26, 27]. In the Penrose hypothesis, speculation and right arguments are interwoven. Therefore, it deserves further analysis. Penrose initially argues as a mathematician that mathematics cannot be traced back to digital machines modelled on the Turing machine. In fact, mathematical proofs must distinguish degrees of computability beyond the Turing machine [28]. The mentioned oracle machines and analog networks are examples of this.

But, Penrose goes one step further and wants to use quantum physics to explain the phenomenon of human consciousness. He describes the complex coordination of many partial states in the brain, which is necessary for conscious thinking, by a quantum physical superposition. This corresponds to quantum parallelism in a quantum computer. The "reading" of results takes place in the quantum computer by a reduction of the superposition. Since this reduction is quantum-physically in principle unpredictable or non-algorithmic ("random"), Penrose also tries to justify the creativity and superiority of human thought over the deterministic computer. In contrast, algorithmic processes like in a computer do not require consciousness.

This also corresponds to our intuitive idea that routine activities take place unconsciously.

The neurobiological speculation of Penrose, according to which the state of consciousness associated with superposition can be explained in the so-called microtubules of the brain, is particularly controversial. Microtubules are tiny protein tubes in the cytoskeleton of cells. A conscious event occurs when a superposition occurs in many microtubules distributed over the entire brain. This would presuppose that the microtubules also contain the suitable medium with which this quantum effect can be maintained.

However, quantum physical superpositions in nature are of such short duration that they decay before they can influence neuronal processes. The brain is probably much too warm for superpositions that are produced in the laboratory at very low temperatures. It is undisputed that quantum effects can also have an effect on the molecular and cellular level. For example, quantum chemistry describes quantum processes during the ejection of transmitter molecules that are involved in the occurrence of action potentials. However, the maintenance of a superposition, which should be associated with the appearance of a thought, is much greater than the measured quantum effects in the brain.

Quantum physics is the basis for the evolution of nature. At the beginning of the universe, there was a quantum vacuum from which elementary particles and atoms developed. This basic layer of nature can only be described with the laws of quantum physics. Depending on their size, the resulting molecular structures are at the interface of quantum chemistry and classical physics. Biological systems up to and including the metabolism in brains can be explained within the framework of chemistry and classical physics. Classical physics can be approximatively embedded into quantum physics if we consider, for example, "slow" velocities (relative to the speed of light), "large" systems (relative to elementary particles), and "weak" gravity (relative to the attraction of black holes).

It seems that microsystems, through the peculiarity of their (non-linear) interaction, lead to the formation of new macroscopic structures—from elementary particles, atoms, and molecules to organs and brains. In the opposite direction, organ states can be explained by cellular interactions, cell states by molecular interactions, molecular states by atomic interactions, etc. In Sect. 10.1, the principle of local activity in complex dynamic systems was introduced in order to mathematically describe the formation of complex structures in nature. It is remarkable that macro states of a complex system cannot be reduced to the individual micro states—from the superposition of quantum systems to the life of cells and organisms.

All previous measurements and observations indicate that the formation of new structures and states in the brain can also be explained "in layers": Quantum mechanical interactions of elementary particles generate quantum chemical states in synapses, whose molecular interaction leads to interconnection patterns of neuronal networks that are connected to cognitive states of the brain. States of consciousness are therefore, as has already been explained in Sect. 7.3, not unsolvable "miracles". Doctors are already using their knowledge of the underlying neuronal circuit patterns to sedate patients step by step during operations or to put them under anaesthesia or in a coma.

While in machine learning, however, the development of perception from neuronal circuit patterns is technically generated, the previous knowledge about states of consciousness—at least as we know it from humans and higher living beings—is not sufficient to generate consciousness technically: Self-perception of today's robots is only the first step in this direction.

As has been stressed several times in this book, technology has by no means been and will be limited to the simulation of natural intelligent systems. In Sect. 10.1, neuromorphic computer structures were explained, which do not occur in this way in nature, but combine the advantages of neuronal systems of nature with the advantages of technical computer structures. Neural quantum computers are also conceivable, in which the enormous computing speed and storage capacity of quantum computers are combined with neural networks. In the end, it is technically impossible to rule out that the hypothesis of Penrose that states of consciousness in the human brain can be explained by quantum physical superpositions is neurobiologically wrong, but could one day be realized with a quantum physical computer. The first technical challenge is to realize superpositions over a longer period of time than in nature, independent of environmental conditions. But, whether and how they can be connected with states of consciousness is another question.

▶   A basic thesis of this book is that the biological evolution of intelligent systems was only one possibility that more or less coincidentally occurred on this
planet. Within the framework of the laws of logic,
mathematics and physics, other technical developments are quite possible, some of which have already
been implemented. Within the framework of these
laws, the innovation area is in principle open.

Will this lead to epistemological breakthroughs, according to
which previously undecidable and unsolvable problems with
quantum computers become decidable and solvable?

▶   The fundamental undecidability and unsolvability of
problems are based on laws of logic and mathematics. Therefore, even a quantum computer, in principle,
will not solve more than is possible according to the
logical-mathematical computability theory: Problems
that are algorithmically unsolvable and undecidable
remain unsolvable even for quantum computers [29].

For example, the stop problem of a Turing machine is also undecidable for a quantum computer. A further example is the word
problem of group theory, according to which for any two expressions of a symbol group it must be checked whether they can be
converted into each other by predefined transformation rules.
Behind this is a problem that often occurs in practice, whether
expressions in language systems can be traced back to each other
or not.

In computability theory, it has been proven that there is no
algorithm that can make a decision in every case. No quantum
computer will change that either. So even in a civilization with
quantum computers, there will be no machine that can solve all
problems algorithmically. Gödel's and Turing's logical-mathematical limits will therefore remain, even if there are gigantic increases in computing speed and capacity. Every kind of
physical, chemical, biological, and neuromorphic computer will

respect the laws of logic and mathematics—as well as the evolution of nature itself.

In addition to the principle of superposition, another (classically) strange phenomenon of quantum physics states that two spatially distant bodies such as elementary particles can be correlated ("entangled") via a common quantum state, although they do not interact with each other via any mechanism.

▶ In EPR (= Einstein-Podolsky-Rosen) experiments, e.g., photon pairs were analyzed, which fly from a central source in opposite direction to polarized filters. [30]. The correlations of the polarity states are understood as entanglements of the locally separated ("local") photons in a ("non-local") overall state. Based on a correlation, a measurement made on one system now determines the result of a measurement on the other system at the same moment. This cannot be understood for two balls flown apart in classical physics, but is precisely predicted by quantum mechanics for quantum systems and confirmed in the EPR experiment.

Classical information can be transmitted between transmitters and receivers, which can be realized by different physical, chemical and biological carrier systems. However, transmitters and receivers must not be miniaturized in the size range of quantum effects. In the quantum world, the transmitter corresponds to the preparation of a quantum system, the receiver to its measurement. The quantum systems (e.g., elementary particles), which develop from the prepared state of an experiment to measurement, convey information in this sense [31].

▶ Quantum information is understood as the information transmitted by quantum particles from the preparation to the measuring apparatus of a quantum mechanical experiment.

Measurements and observations are carried out on the systems prepared in this way: Repeated experiments can now be carried out on the same experimental setup with preparation of the initial state of a quantum particle and measurement of its final state.

For such a series of measurements, the relative frequencies of the test results are then noted and used as a basis for statistical predictions. If there are gross deviations from the typical behavior of a random sequence, the measurement is considered to have failed.

In quantum physics, entangled quantum states can be used, which allow an instantaneous quantum teleportation of quantum information to remote receivers. This is no contradiction to the theory of relativity, according to which signal transmission is only possible at the speed of light. In fact, it is not an "interaction" between two objects located at different locations. In quantum physics, the EPR correlation produces a single quantum state which is distributed over the space between the two objects.

The problem with quantum teleportation, however, is that the quantum information to be sent is unknown and only decides by chance of a measurement. Therefore, quantum teleportation cannot be used for direct information transmission. In this respect, there is also no conflict with the theory of relativity, according to which no interaction can be faster than light. However, as long as we do not measure, read and observe quantum information, it can be transmitted instantly and in any superposition.

Technically, entangled states have already been realized over kilometers of long distances on Earth. With the statistical limitations mentioned above, quantum teleportation can be realized. The speed of light may not be an effective limit for the transmission of information on the Internet on Earth. In space travel to Mars in a few years' time, however, the delay caused by the speed of light in the transmission of information and control from Earth is already becoming a problem. Therefore, the technical realization of entangled states on a cosmic scale will be a challenge of the future. In chap. 9, it was about communication with intelligent infrastructures on the basis of the Internet of Things at the speed of light. We dare to predict that interstellar space travel will be supported by a quantum Internet of Things, in order to realize communication with the Earth.

How can intelligent infrastructures on a cosmic scale be realized on the basis of a quantum Internet?

10.4   Singularity and Superintelligence?

So far, it has been shown how AI systems outperform people in certain sectors. AI systems can calculate, answer and ask questions faster, recognize and predict relationships, handle larger amounts of data, have a more extensive memory, etc. With all the exponential increases in computing and storage performance, however, nothing changes in the basic principles: undecidable problems remain unsolvable. Nevertheless, at least theoretically, amazing increases in intelligence are conceivable in AI systems that are superior to humans. In this case, we are talking about superintelligence [32]. The following examples serve as an explanation:

Example

1. Fast superintelligence: An AI system can do everything humans can do just faster.

   We know the experience when we are faster than others or others faster than us. In the first case, we experience a stretching of time and become increasingly bored. Otherwise, we feel overwhelmed. It is conceivable that artificial AI systems based on silicon or organic substances, such as carbon in nanoelectronics according to Moore's law, could increase their performance. Neuromorphic systems using neural tissue developed in synthetic biology could also be possible. This tissue could be implanted both in case of damage and for enhancement in living brains. Enhancement of the natural brain is ethically controversial and would also reach medical-biological limits.

   In the case of artificial AI systems, any increase in performance below the speed of light would be conceivable within the framework of physical laws. In terms of speed, the technical possibilities of the quantum computer would go beyond everything that has existed so far, but would not change anything about the logical-mathematical limits and laws of computability.

2. Collective superintelligence: An AI system consists of many subsystems that can perform less than humans. However, the collective AI system is superior to humans.

This strategy of increasing intelligence was already conceived in evolution as swarm intelligence. The collective intelligence, e.g., of a termite population exceeds the possibilities of individual animals by far. Only the interaction of all animals creates the refined termite structures. The communication between the insects uses chemical codes. The similarity with the neurochemically communicating neurons in the brain is unmistakable. Here, too, the individual neuron is "stupid", but the collective interaction creates intelligent problem solving. Swarm intelligence is also used in the robotics technology (see Sect. 8.3).

Transferred to humanity, the interaction of many people creates a collective intelligence of humanity that is far superior to the individual [33]. For generations, this collective intelligence of humanity has been passed on through libraries and educational systems, increased exponentially through computers, databases. and the Internet, and is increasingly becoming independent of the individual human being. We observe intelligence enhancements that are no longer due to individual contributions, but to synergetic effects of their cooperation. These include the intelligent infrastructures referred to in the Sect. 9.2 above. Increasingly, automated decisions are also being made by these collective systems. This does not require any consciousness as we humans do. Collective super intelligence, which is superior to man, becomes highly probable and is foreseeable. It is a great challenge to steer this intelligence and to make us human servants.

3. New superintelligence: An AI system has new intellectual abilities that people don't have at all.

This strategy has also been laid out in the history of technology. Inventors and engineers find intelligent solutions to problems that were by no means predetermined in evolution. A classic example is flying, which humans do not master by nature, but which they realize differently with jet planes than birds with wings. Meanwhile, artificial skin sensors are being developed for robots that do not only register temperature and pressure, but also radiation and chemical signals. But, new intellectual abilities are also conceivable: Robots become feasible that use the entire Internet as their memory and apply machine learning algorithms to Big Data at lightning speed. Finally, human brains could be complemented by neuromorphic circuits with novel intelligent capabilities (e.g., an adapter for automatic language learning).

As has been shown many times in this book, the architecture of the human brain is different from the digital technology of our computers. In contrast to targeted and conscious optimizations of technology, which took place in a short period of time, brain architecture evolved more or less randomly over millions of years under changing conditions and requirements. Biological nerve cells developed over long periods of time from cells that first casually, then increasingly, generated nerve signals in order to finally specialize in the generation of action potentials for control and regulation tasks. This led to a highly sophisticated neurochemical signal processing with synapses and ion channels, which made our intellectual abilities possible.

On the other hand, biological neurons can only fire very slowly when compared to a modern microprocessor. This slowness was compensated in the brain by an increased expansion of parallel signal processing and led to the enormous network density of the human brain. With these complex networks and learning algorithm, the brain's pattern recognition, which is so crucial for the survival of animals, including humans, became possible. In contrast, signals are processed sequentially in the von Neumann architecture of a computer. The technology relies on the enormous speed of signal processing, which is possible with silicon hardware according to Moore's law.

From a logical-mathematical point of view, both approaches are equivalent—a (von Neumann) computer and a Turing machine respectively and a (recurrent) neural network (with rational weights). With super intelligence, the advantages of one approach could be used to offset the disadvantages of the other approach. This is how the material of technical microprocessors, transistors, and memristors is more stable and resilient than biological neurons, axons and synapses. In case of defects, they can be replaced like used bulbs. Biological tissue, on the other hand, is subject to ageing processes and pathological changes (e.g., tumour formation), which we have not understood for a long time. Therefore, technical brain networks are technically conceivable that could exchange much faster and more resilient signals than biological neurons and synapses.

Another example: A biological short-term memory has the advantage of short access times, but the disadvantage of small storage capacity. The in-memory technology of Big Data databases already combine the advantage of a biological short-term memory with enormous data memories.

In addition to the hardware or wetware, optimized software applications are also conceivable. Considering how hard it is for a human brain to learn and store little data. Errors and redundancies are typical of biological brains. A computer, on the other hand, transmits enormous amounts of data exactly at the touch of a button and duplicates them in any number for other computers. On the other hand, dealing with mistakes, redundancies and noise led to an "eye for the essentials" that does not get lost in details. The effective evaluation of patterns and the discovery of overall contexts characterizes human intelligence.

Therefore, the advantages of human intelligence include the neuronal areas, which enable rapid intuitive assessments and decisions, i.e. procedural learning. But it is technically by no means impossible to develop algorithms specialized on it in a superintelligence as well.

Which scenarios are conceivable in which a superintelligence could develop? In Fig. 10.6, a takeoff of a super intelligence in several steps is assumed. The first stage would be a basic human level at which a human brain can be fully simulated by an individual AI system. This could be an artificial neuromorphic system. A computer that is logically and mathematically equivalent

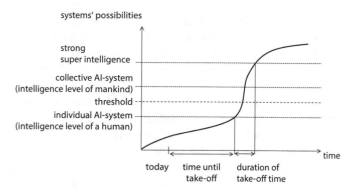

Fig. 10.6 Takeoff of SuperIntelligence [32]

to the one Turing had originally imagined is by no means out of the question.

In principle, all living brains could then be replaced by such an AI system. This would create a collective AI system equivalent to the collective intelligence of humanity. On the way there, subsystems would have to develop (e.g., intelligent infrastructures), which increasingly make independent decisions and set themselves goals, because only then can they determine consequences better than humans. In this way, a threshold would be crossed that would eventually lead to a superintelligence.

▶ **Definition**
Superintelligence is an AI system that is superior to individual and collective human intelligence.

An AI system with the ability to improve itself is called "Seed AI" (germ or seminal intelligence).

▶   **Important**
   However, even a superintelligence is subject to
   a) the logical-mathematical laws and evidence of
      predictability, decidability, and complexity (see
      Sects. 3.4, 5.2 and 10.2),
   b) the physical laws.

Logic, mathematics, and physics remain the legal framework of every technological development. Therefore, we need interdisciplinary foundational research so that the algorithms do not get out of hand.

The question remains whether this superintelligence will arrive at a certain point in time or whether it will develop in a continuous long-term process. The proponents of a selective event argue: Since building a superintelligent machine is one of the capabilities of a superintelligent machine, an explosive development of improving superintelligences could occur. As early as 1965, the statistician I. J. Good, who had been working during the 2nd world war with Turing, predicted: "The ultraintelligent machine is thus the last invention that man has to make" [34].

▶ Technological singularity means the time at which a superintelligence emerges.

In 1993, the mathematician V. Vinge published an article entitled "Technological Singularity" in which he linked the end of the era of mankind with a technological singularity [35]. The computer scientist and author R. Kurzweil attributes singularity to exponentially growing technologies [36]. This includes not only computing capacity according to Moore's law, but also nano and sensor technology, as well as genetic, neuro and synthetic biology technologies to generate new forms of life. The convergence of these exponential technologies opens up future perspectives that are associated with dramatic possibilities for human change as a biological species: Why should people with this potential accept the aging process and death as "natural"? The actual evolution on Earth was itself only a branch of development in a multitude of possibilities within the framework of the laws of evolution (cf. chap. 6), which was realized under more or less random conditions.

Authors like Kurzweil at the Singularity University in Silicon Valley propagate a transhumanism in which all diseases are defeated and the social problems of mankind can be solved. The American dream according to the slogan "Yes, we can" is combined with the assumption of unlimited technological potentials. Successful entrepreneurs like, e.g., Bill Gates want to defeat the diseases of mankind with their capital and modern medicine technology. Successful researchers like Crai Venter explain the transformation of life into a profitable business model.

▶ Transhumanism wants to overcome the limits of the human organism through technology. Beyond technological singularity, a superintelligence will control human development.

What years ago was still considered scientifically dubious phantasms of science fiction authors, increasingly takes on realistic traits. In the end, however, there remains the question of how super-intelligence can be controlled at all following singularity as a service for human beings. In order to transform AI into a successful business model, influential robotics laboratories are also allying themselves with military technology. A military arms race of AI weapon systems could initiate the development spiral into a superintelligence that is no longer interested in the general welfare of mankind.

Decades ago, molecular biologists and genetic engineers warned against the misuse of their knowledge and know-how. In view of the increasing autonomy of AI systems, renowned scientists and technology entrepreneurs are now warning against a digital arms race of AI and drawing parallels with nuclear weapons. The unlimited energy of nuclear fission corresponds to the unlimited increase o a superintelligence—in both cases uncontrollable.

Artificial Intelligence will converge with other developments of technology. Technological developments of the future can be assessed scientifically seriously with the following questions:

Fragen

- Which technologies are possible today?
- For which technologies do we already have the technical possibilities to build them in the next years (e.g., energy transport by laser beams, electric vehicles)?

- Which technologies are physically conceivable, but their technical realization still fails due to many difficulties (e.g., fusion reactor, rocket propulsion by fusion)?
- Which technologies are physically conceivable, but without the currently foreseeable technical possibility of realization (e.g., rocket propulsion by antimatter, vehicle propulsion with superconductors)?
- What role will nanotechnology and robotics play in space? How do the developmental stages of space technology depend on the developmental stages of human civilization?

It is remarkable that the computer pioneer K. Zuse was already planning concrete projects on the future of AI in space in the 1960s. If the universe, according to Zuse, is a computable cellular automaton, then automata should also be used for its settlement. At the beginning of the theory of cellular automata, there was the question of how automata should reproduce themselves analogously to living organisms. Mathematically, the problem was solved by J. von Neumann with a universal cellular automaton (see Sect. 6.2). The engineer and inventor Zuse, however, was concerned with the technical problem of building a corresponding robot. In the early 1970s, Zuse started his project of self-reproducing systems with the construction of the SRS72 assembly line, which would make a copy of itself with supplied workpieces. His restored assembly line is now in the Deutsches Museum in Munich [37].

Background Information
Zuse associated this with the vision of a technical germ cell that can reproduce itself with system-internal data storage and data processing by resorting to available raw materials in order to grow into a complex system like a biological organism. With such germ cells, according to Zuse, human civilization could spread into space: From germ cells on one planet, intelligent robot factories are created, which in turn produce germ cells that are shot at other planets in other star systems in order to repeat the self-reproduction process there.

In 1980, American physicists described these scenarios as "von Neumann probes". Unlike Zuse, von Neumann never mentioned such a space project.

New opportunities for materials research, such as those opened up by nanotechnology, will certainly be of central importance for technical self-reproduction. At first, technologies that reproduce themselves and act more or less autonomously with humans will be integrated into socio-technical systems. The Internet of Things and Industry 4.0 are first steps in this direction. The "human factor" will be a central challenge in this development, in order to adequately take into account its organic, psychological, and intellectual preconditions.

The state of the art in the life sciences will have to be taken into account. According to current knowledge in systems biology, it cannot be ruled out in principle that life can be prolonged at will. Scholars are still arguing whether death is more genetically determined and/or through an evolutionary fitness program to give the population greater overall chances of survival. Mathematical models of population dynamics provide amazing evidence of this.

This example directly illustrates the social and societal consequences of such long-term future scenarios. Today's technically and economically highly developed societies already have enough difficulties to cope with the demographic change of a society whose people stay fit longer in old age. Finally, the political questions: How are conflicts resolved in such societies? What forms of social and political organization will be appropriate under these conditions?

In this book, it has been shown that the concept of information is the universal category, with which not only technical, but also social, economic, and societal changes can be captured. Therefore, the American astrophysicist C. Sagan (1934–1996) proposed a scale that measures civilizations according to the state of data processing [38]. His scale runs from the letters A to Z, which correspond to increasing data capacities:

Background Information

A Type-A civilization can only handle a million bits. This would be a stage of development in which only spoken language but no written language with documents can be used. One thinks of native peoples, as they were discovered, e. g., in the Amazon region.

An ancient high culture such as Greece, with its traditional written documents, has an estimated size of one billion bits and corresponds in Sagan's scale to a Type-C civilization.

Sagan's assessment of current civilization was before the big data age. With Big Data we are on our way from peta (10^{15})-byte age to the exa (10^{18})-, zetta (10^{21})- and botta (10^{24}) byte age.

According to the ideas of Silicon Valley, after the singularity, it's really happening with a superintelligence. The spread of the human species in the universe will also require superintelligence related to today's standards. The other side of the coin is the enormous demand for energy.

Every kind of civilization depends on its energy consumption. The Russian astrophysicist N. Kardashev had already considered in the 1960s, how the state of development of future civilizations could be classified according to the possibilities of their energy consumption [39]. Then, a quantitative scale with measurable quantities is obtained. Kardashev distinguished three types of civilizations:

Background Information

Type-1 civilization controls the energy of its planet. The consumable energy of a planet is determined by the fraction of the incident light of its sun. In the case of Earth. we can estimate a size of approx. 10^{17} W. This does not only refer to solar energy, which is now generated by solar power and photovoltaics. Fossil fuels are solar energy stored in dead plants. Wind, weather, and ocean currents are also only possible through solar energy. A civilization of this type dominates all these forms of energy. This seems utopian for mankind at present, but physically not impossible.

Therefore, mankind is still a Type-0 civilization with energy consumption less than 10^{17} Watts. Quantitative fine scaling can be specified for this purpose. It all started with a 0.2 HP (horse power) civilization that could only build on the physical strength of humans. This is the time of hunters and gatherers before animal breeding began. The 1 HP civilization, which was supported by horsepower, reaches into the stagecoach age. It was not until the motorization of steam engines at the beginning of the 19th century and finally combustion engines at the end of the 19th century that energy consumption changed exponentially—on the basis of coal and oil, finally nuclear energy. Meanwhile we are tapping into the energy on this Earth in all possible forms of storage. But, from a control of wind and weather, the mankind is still far away, although the possibility is not impossible with use of appropriate laws. The current percentage estimation according to the degree of energy use is therefore a type 0.7 civilization for mankind.

In the mathematical theory of plasma physics, we have already packed the fusion energy of the sun into formulas. However, the fusion reactor is still a long way off. After Cardashev, this would be the first step in the direction of a Type-Two civilization: It controls the energy of the sun, thus approx. 10^{27} Watts. This does not only mean solar cells that passively collect solar energy. The American physicist F. Dyson describes how such a civilization constructs a gigantic sphere around its home star to absorb all its radiation.

A Type 3 Civilization is galactic and consumes the energy of billions of stars in the order of 10^{37} Watts.

The Kardashev scale we can only be illustrated so far in pictures, how they are derived from science fiction literature. Type-1 civilization would be the world of Flash Gordon, because all planetary energy sources can be used there. Even wind and weather are then completely controllable. Type-2 civilization is the planetary federation in Star Trek, which has already colonized a hundred nearby stars. After all, the empire in the movie "Star Wars" corresponds to a type-3 civilization: large parts of a galaxy with billions of stars are used. Under these conditions, superintelligence is spreading in the universe …

10.5 Technology Design: AI as a Service System of Mankind

Since the future prognoses of the Singularity followers argue with measurements such as computing and storage capacity, decrease of size, price reduction, efficiency increase etc., the digital future of mankind seems to be determined by corresponding exponential curves. In this case, the exact time of singularity would only be a question of fine-tuning of the presumed conditions.

This digital determinism is extremely problematic, as previous technological developments have by no means been deterministic. New innovation impulses often gave developments unforeseen directions. At the beginning of the 1950s, computer pioneers relied on a few mainframe computers. But, then came Bill Gates with his many small personal computers (PC). Even the Internet, as the basis for worldwide communication, was initially not on the screen when military communication networks were set up to secure command structures in the event of a nuclear strike. The exponential success of smartphones and associated companies was also not planned for the long term. Nobody knows today which developmental boosts can be expected in the coming decades and which trend reversals they could initiate.

Technology development has a certain similarity with biological evolution. Innovation plays the role of mutations. Markets act as selections. Social framework influence trend developments such as ecological conditions influence evolution. But, during millions of years, the algorithms of evolution were "blind". Humans are (still) conscious of the development of technology. At least during short periods of time, they can control and influence trends in a targeted manner.

Conversely, future models influence human goals and desires and influence future development through this feedback of human consciousness: It's called the normative power of the factual. In this way, Silicon Valley's idea of singularity can generate believing supporters of transhumanism. If they are among the top performers of key companies and research centers, the end result could be a self-reinforcement that actually makes the predicted become reality ("self-fulfilling prophecy").

Evolution already shows, however, that the developmental possibilities within the framework of the laws of nature are open. Only some branches of the possible ramifications were realized. Therefore, the rule is

▶   **Important**
The future is open within the framework of the laws of nature. One therefore also speaks of "futures" instead of "a" future [40, 41].

Technical development can be influenced under changing technical, economic, ecological, and social conditions. This is what we call technical design.

While Big Data analyze the future with powerful algorithms, the older scenario and Delphi techniques try to gain qualitative insights into the future. This is intended to test the scope for technology design.

The scenario techniques rely on a qualitative understanding of events. Instead of supercomputers and data, it is therefore a matter of human knowledge and understanding by experts, which is to be evaluated for possible scenarios of the future [42]. These approaches do not aim at a determined future, but consider a potential of possibilities ("futures") that depends on the experiences, ideas, and intuitions of the selected experts.

▶ Definition
Scenarios describe future situations and states that are assumed as hypotheses and from which causal and logical consequences are drawn.

These consequences allow an assessment of alternative future scenarios as more or less desirable.

The starting situation is the present and the past, insofar as they are accessible through empirical data analysis.

From this, a trend scenario is determined that is extrapolated into the future under the assumption of constant constraints.

Background Information
However, an assumed variation of the ancillary conditions would result in alternative possibilities of scenarios that deviate further and further from the trend scenario with increasing distance from the present.

In intuitive terms, a kind of funnel is formed which, starting from the present, opens itself ever further around the time axis of the trend scenario. Positive and negative extreme scenarios can be distinguished at the margins.

An example is provided by future scenarios of energy development which, based on a continuation of the present conditions, play through different scenarios of varying political decisions.

During the elaboration of the scenario development, phases are distinguished which range from the initial and task analysis to factor and trend analysis and the derivation of consequences, finally their evaluations and interpretations.

The Delphi method is used as the valuation method for future developments [42]. The legendary oracle of Delphi is now being replaced by experts who, on the basis of their knowledge, are to identify and evaluate trend and future models. This instrument was used by ministries and science organizations, for example, as a decision-making aid when investments for future innovations were to be justified.

▶ **Definition**
In a Delphi procedure, in a first round, a questionnaire is presented. In a second round, the participants are informed about the evaluations of the other experts, in order to initiate further evaluation rounds step by step by feedback with other evaluations, until a consensus or stable alternatives result at the end.

The Delphi rounds are concluded with a roadmap, which recommends a strategy for action or an implementation plan for a project to the client.

Scenario and Delphi techniques do not make futures completely computable, but plausible. In Turing's terminology, experts with their knowledge and intuitions can be understood as oracle machines whose results can be combined with computable and provable arguments.

A weak point is the selection of experts. As long as one moves with his trend evaluations in a limited discipline, this may still be unproblematic. However, when it comes to the future assessment of socio-technical systems, the initial situation becomes much more difficult: If you only ask engineers to build an energy plant or an airport, you will only experience the engineer's point of view. Those who only ask social scientists will only hear their assessments based on social science methods.

In addition, there is the public concerned. Here an evaluation and communication process is emerging, which must not only convey interdisciplinary knowledge, but also opinions and attitudes. Socio-technical systems are complex, their realization under the conditions of democracy even more complex. In the end, however, robust decisions must be possible in order to assess future risks.

Example

Technology design using the example of an intelligent infrastructure.

It is crucial that computer networks are integrated into society's infrastructures and that social, economic, and ecological factors are taken into account. Thus, AI-supported socio-technical systems make it possible to provide services to people. They are networked with their environment (e.g., Internet), should be robust against disturbances, adapt, and react sensitively to changes (resilience). Applications can already be found in the workplace, in the household, in geriatric and nursing care, in transport systems, and in aviation.

Intelligent infrastructures are complex systems that have to integrate technically different domains [43]. The infrastructures of an intelligent factory, an intelligent health center, or an intelligent transport system are to be recorded in a common software. The software of a computer distinguishes between the user level and the middleware with translation programs into the machine language. An intelligent infrastructure such as a city or an airport is understood as a virtual machine. First, there is the level of integrated customer and usage processes at which customers and users communicate and interact with the system. Interoperability is visible to the user. Such services will be integrated at the lower level according to usage needs. Then the domain-specific architectures of a transport system, the health care system, and an industrial plant are accessed.

In concrete terms, we can imagine a city administration that is to be represented in a common software and that has to take into account the urban transport system, health care with various authorities, and industrial facilities of the municipal energy supply and mill incineration plants. The provided interoperability of the services thus receives concrete applications (semantic interoperability).

▶   **Important**
    The technical design of information infrastructures
    requires interdisciplinary cooperation of the techni-
    cal, natural, social and human sciences with physics,
    mechanical engineering, electrical engineering, com-
    puter science, but also cognitive psychology, commu-
    nication science, sociology, and philosophy.

    What is required are models of perception, inte-
    gration, knowledge, thinking and problem solving,
    but also system and network models of sociology and
    philosophy of technology.

    The goal is integrative human factor engineering
    of information infrastructures.

Human-centered engineering aims at integrated hybrid system
and architecture concepts for distributed analog/digital control,
human-technology interaction and integrated action models, socio-
technical networks, and interaction models. This requires the step-
by-step development of reference architectures, domain models,
and application platforms of individual disciplines as prerequisites
for conscious situational and context perception, interpretation,
process integration, and reliable action and control of the systems.

Human factors in information infrastructures must be
researched in an interdisciplinary manner—from classical questions of ergonomics, the integration of adaptive and adaptable
structures in the workflow and the corresponding effects of
traceability to problems of adapting social behavior under the
influence of the use of appropriate systems. Simple, robust, and
intuitive human-machine interaction is recommended, despite
multifunctional and complex services and action options.

Example
Criteria of technology design

Sensitivity is required for increasing loss of control in
open social environments with complex networked and autonomously interacting systems and actors, reliability and trust
of the systems with regard to safety, IT security, and privacy.
The benchmarks are as follows:

- performance and energy efficiency (environment),
- know-how protection in open value chains,
- assessment and evaluation of uncertain and distributed risks,
- appropriate and fair conduct in the event of conflicts of objectives between different subsystems, binding domain and quality models, rules and policies to be negotiated (e.g., compliance).

Intelligent infrastructures develop against the background of a changing society: they also change the structures of democracies. Digital communication enables citizens to obtain information more quickly. Changes in society that could entail new socio-technical systems lead to a significant increase in attention by civil society organizations, NGOs, and the public. Due to real-time information, higher reactivity in increasing network density and associated cascade effects, new liquid (non-rigid and "liquefied") forms of democracy are emerging. Better and faster information prompts citizens to demand greater participation in decisions on the introduction of socio-technical systems.

▶   Technology design is not only a task for experts, but also involves society.

Greater participation by civil society responds to the demand for participatory democracy. To this end, technical solutions must include ecological, economic, and social dimensions. In this case,we talk about sustainable innovations. Despite greater participation, large-scale socio-technical projects should remain feasible in order not to endanger the innovation location. Sustainable innovations should therefore also be robust. Sustainable and robust innovations are what make a society's future viability possible in the first place.

In global digitization, information and knowledge multiplication, sustainable information infrastructures are the prerequisite for society's innovation potential. This requires the creation of integrative research and teaching centers in which engineering and natural sciences, together with humanities and social sciences, prepare for the challenges of socio-technical systems [44].

In these interdisciplinary research clusters, the university of tomorrow is emerging. They are transverse to the traditional faculty distinctions of engineering, natural sciences, social sciences, and the humanities. This is why we speak of a matrix structure: the disciplines are understood as matrix lines. The matrix columns are the integrative research projects that pick up different research elements of the disciplines. The TU Munich has founded the Munich Center for Technology in Society (MCTS) as part of the Excellence Initiative 2012. In 1998 at the University of Augsburg, the Institute for Interdisciplinary Computer Science (I3) was already founded, in order to analyze the societal impact of the (then) Internet.

Behind this is the fundamental insight that science does not work independently of society. Without considering social structures and social processes, hardly any innovation in engineering and natural sciences (especially AI research) can be successful.

Questions

How could intelligent cities (smart cities) be created without any knowledge about the future coexistence in the cities?

How should researchers develop intelligent food and supply chains for the world's growing population without considering the situation in developing countries?

How could robots help old people without taking their needs into account?

How should large-scale technology projects such as intelligent energy networks be integrated into society without taking into account the associated social, economic, and ecological factors?

Not only applied research, but also foundational research is confronted with questions that cannot be answered without social sciences and the humanities:

What are the criteria we use for our research?

How can science work beyond our common understanding?

How do we learn from failed approaches?

▶ **Important**

Questions of humanities and social sciences must be addressed right from the start in the design of technology and not only in a subsequent "add-on" that comes into play when the technology has already created facts.

The interactions between science, technolog, and society must be examined from three perspectives—knowledge, evaluation, and communication:

Science & Technology Studies (STS): Social scientists and humanistic scholars research the social aspects of science and technology—including philosophers, historians, sociologists, political scientists, and psychologists.

Ethics & Responsibility: Economic and medical ethicists, environmental and technical ethicists evaluate research and development.

Media & Science: Communication and media scientists examine how research and society can exchange ideas.

In technology design, human scientists concentrate on the empirical investigation of concrete problems. For this purpose, labs should be established that meet the following criteria:

1. Research projects are interdisciplinary in natural, social, and engineering sciences ("interdisciplinarity").
2. They are project-oriented, i.e. they develop questions of ethical and social science from concrete projects ("project orientation", "bottom up").
3. They are designed for public dialogue ("transparency", "glass laboratory"). Therefore, these laboratories are already open for public discussion during ongoing research. The joint findings should also serve as a basis for policy decisions.

In an increasingly informed society, the call for participation in decision-making on infrastructure and technology projects is becoming louder and louder. The previous response of the rule of law were planning approval procedures in which the phase transitions from the preparation of the plan by the project developer to the consultation procedure, public interpretation, discussion, forwarding of the result of the consultation up to the planning approval decision were legally precisely defined.

However, the participation of citizens and authorities is often declared as a "hearing" in a manner that appears to be in the hands of the authorities. A so-called "preclusion effect" excludes any kind of objection after expiry of the preclusion period. In this case, learning processes are not possible, although technical, social, and economic conditions can change. It is a "linear" legitimation procedure that must take account of a changed complex world.

To what extent is participation possible without gambling away the decision-making capacity and sustainability of a society? The rules of the game between citizen participation, technical-scientific competence (research institutes, universities etc.), the parliaments as democratically legitimized decision-makers, the judiciary and the executive must be rethought. The technical-economic-ecological development is changing political structures.

▶ **Important**
The aim must be for future generations of engineers, computer scientists, and scientists to take the link with society for granted as part of their work.

To this end, students in all subjects must be sensitized.

The technical design of the human-machine relationship in AI research is only possible under consideration of human scientific factors.

The big questions of the future of artificial intelligence can only be answered in an interdisciplinary way.

Each development step should be examined at both the technical and the organizational level in order to discuss social implications and challenges in dialogue with society and to draw consequences.

With this strategy, we could avoid to oversleep the "singularity". Otherwise, one morning, we wake up in the brave new world of a superintelligence with its transhumanism.

References

1. Mainzer K (2005) Symmetry and complexity. The spirit and beauty of nonlinear science. World Scientific, Singapore
2. Mainzer K (2005) Thinking in complexity. The computational dynamics of matter, mind, and mankind, 5th edn. Springer, Berlin
3. Mainzer K, Chua L (2013) Local activity principle. Imperial College Press, London
4. Banerjee R, Chakrabarti BK (2008) Models of brain and mind. Physical, computational, and psychological approaches. Progress in brain research. Elsevier Science, Amsterdam
5. Chua L (1971) Memristor: the missing circuit element. IEEE Transaction on circuit Theor 18(5):507–519
6. Chua L (2014) If it's pinched it's a memristor. Semicond Sci Technol 29(10):104001–104002
7. Sah MP, Kim H, Chua LO (2014) Brains are made of memristors. IEEE Circuits Syst Mag 14(1):12–36
8. Williams RS (2008) How we found the missing memristor. IEEE Spectr 45(12):28–35
9. Tetzlaff R (ed) (2014) Memristors and memristive systems. Springer, Berlin, p 14 (according to Fig. 1.5)
10. Siegelmann HT, Sontag ED (1995) On the computational power of neural nets. J Comput Syst Sci 50:132–150
11. Siegelmann HT, Sontag ED (1994) Analog computation via neural networks. Theoret Comput Sci 131:331–360
12. Blum L, Shub M, Smale S (1989) On a theory of computation and complexity over the real numbers: NP-completeness, recursive functions and universal machines. Bull Am Math Soc 21(1):1–46
13. Turing AM (1939) Systems of logic based on ordinals. Proc London Math Soc 2:161–228
14. Feferman S (2006) Turing's thesis. Notices Am Math Soc 53(10):1200–1206
15. Mainzer K (1973) Mathematical constructivism. University of Münster, Diss
16. Bennett CH (1995) Logical depth and physical complexity. In: Herken R (ed) The universal turing machine. A half-century survey. Springer, Vienna, pp 227–235

17. Brooks RA (2005) Human machines. Campus Non-fiction, Frankfurt a. M.
18. Wagman M (1996) Human intellect and cognitive science. Towards a general unified theory of intelligence. Greenwood, Westport Conn
19. Wagman M (1995) The science of cognition. Theory and research in psychology and artificial intelligence. Greenwood, Wetstport Conn
20. Mainzer K (2016) Information: Algorithm, Probability Complexity, Quantum World
21. Audretsch J, Mainzer K (eds) (1996) How many lives does Schrödinger's cat have? On the physics and philosophy of the quantum world, 2nd edn. Springer, Heidelberg
22. Feynman RP (1982) Simulating physics with computers. Internal J Theor Physics 21:467–488
23. German D, Eckert A (2000) Concepts of quantum computation. In: Bouwmeester D, Ekert A, Zeilinger A (eds) The physics of quantum information, quantum cryptography, quantum teleportation, quantum computation. Springer, Berlin (Chapter 4)
24. Watrous J (1995) On one-dimensional quantum cellular automata. In: Proceedings of the 36th annual symposium on foundations of computer science. IEEE Computer Society Press, Milwaukee (Wisconsin)
25. Mainzer K, Chua L (2011) The universe as automaton. From simplicity and symmetry to complexity. Springer, Berlin (Chapter 7)
26. Penrose R (2001) Computer thinking: the debate on artificial intelligence, consciousness, and the laws of physics. Springer, Heidelberg (Chap. 10)
27. Penrose R (1995) Shadow of the spirit: ways to a new physics of consciousness. Springer, Heidelberg
28. Mainzer K (2018) The digital and the real world. Computational foundations of mathematics, science, technology, and philosophy. World Scientific, Singapore
29. German D (1985) Quantum theory, the church-turing principle and the universal quantum computer. Proc R Soc London A 400:97–117
30. Mainzer K (1996) Natural philosophy and quantum mechanics. In: Audretsch J, Mainzer K (eds) How many lives does Schrödinger's cat have? On the physics and philosophy of quantum mechanics, 2nd edn. Springer, Heidelberg, pp 245–299
31. Keyl M (2002) Fundamentals of quantum information theory. Phys Rep 369:431–454 (A Review Section of Physics Letters)
32. Bostrom N (2014) Superintelligence. Scenarios of an upcoming revolution. Oxford University Press
33. Shanahan M (2010) Embodiment and the Inner Life. Cognition and Consciousness in the Space of Possible Minds. Oxford University Press, New York

34. Good IJ (1965) Speculations concerning the first ultraintelligent machine. In: Alt FL, Robinoff M (eds) Advances in computers. Academic Press, New York, pp 31–88

35. Vinge V (1993) The coming technological singularity: How to survive in the post-human era. Vision-21: Interdisciplinary Science and Engineering in the Era of Cyberspace NASA Conference Publication 10(129), 11–22 (NASA Lewis Research Center)

36. Kurzweil R (2005) The Singularity is near. When humans transcend biology. Viking, New York

37. Marshmallow N (2011) A machine builds a machine builds a machine builds a machine. Cult Technol 1:48–51

38. Slovskij IS, Sagan C (1966) Intelligent Life in the Universe. Holden-Day, San Francisco

39. Kardashev NS (1964) Transmission of information by extraterrestrial civilizations. Soviet Astron 8(2):217–221

40. acatech (German Academy of Science and Technology) (2012) Technology future. Think ahead—create—evaluate. Springer, Berlin

41. Wilms F (2005) Scenario technique. Dealing with the future. Haupt, Bern

42. Häder M (ed) (2002) Delphi interviews. A workbook. Springer VS, Wiesbaden

43. Geisberger E, Broy M (eds) (2012) Agenda CPS. Acatech study. Springer, Berlin

44. Mainzer K (2012) From interdisciplinary to integrative research. BBAW 28:26–30

How Safe Is Artificial Intelligence?

11

11.1 Neural Networks Are a Black Box

Machine learning dramatically changes our civilization. We rely more and more on efficient algorithms, because otherwise the complexity of our civilizing infrastructure would not be manageable: Our brains are too slow and hopelessly overwhelmed by the amount of data we have to deal with. But how secure are AI algorithms? In practical applications, learning algorithms refer to models of neural networks, which themselves are extremely complex. They are fed and trained with huge amounts of data. The number of necessary parameters explodes exponentially. Nobody knows exactly what happens in these "black boxes" in detail. A statistical trial-and-error procedure often remains. But how should questions of responsibility be decided in, e.g., autonomous driving or in medicine, if the methodological basics remain dark?

▶ In machine learning with neural networks we need more explanation (explainability) and attribution (accountability) of causes and effects in order to be able to decide ethical and legal questions of responsibility!

© Springer-Verlag GmbH Germany, part of Springer Nature 2020   243
K. Mainzer, *Artificial intelligence – When do machines take over?*, Technik im Fokus,
https://doi.org/10.1007/978-3-662-59717-0_11

In statistical learning, dependencies and correlations are to be derived from observed data by algorithms. To this end, we can imagine a scientific experiment in which corresponding results (outputs) follow in a series of changed conditions (inputs). In medicine, it could be a patient who reacts to medication in a certain way. We assume that the corresponding pairs of input and output data are generated independently by the same unknown random experiment. Statistically, it is therefore said that the finite sequence of observational data $(x_1, y_1), \ldots, (x_n, y_n)$ with Inputs x_i and outputs y_i $(i = 1, \ldots, n)$ is realized by random variables $(X_1, Y_1), \ldots, (X_n, Y_n)$ which have an unknown probability distribution $P_{X,Y}$.

Algorithms should now be able to derive properties of the probability distribution $P_{X,Y}$. An example would be the expectation probability with which a corresponding output occurs for a given input. However, it can also be a classification task: A set of data is to be divided into two classes. How likely is it that an element of the data set (input) belongs to one class or another (output)? We are also talking about binary pattern recognition in this case.

Background Information

When recognizing a binary pattern, the data of a data set $\mathcal{X}$ is distributed to two possible classes, which are designated with $+1$ or -1. This assignment is described by a function $f : \mathcal{X} \rightarrow \mathcal{Y}$ with $\mathcal{Y} = \{+1, -1\}$. Statistical learning of a binary pattern involves determining from a class of functions $\mathcal{F}$ the assignment f for which the error deviation or expected error is minimal. We're also talking about the risk minimization of statistical learning [1]:

$$R[f] = \int \frac{1}{2} |f(x) - y| \, dP_{X,Y}(x, y)$$

However, since the probability distribution $P_{X,Y}$ for all values is unknown, this formula and thus the pattern recognition sought cannot be calculated with minimal error deviation. We only have the finitely many empirically observed classifications $(x_1, y_1), \ldots, (x_n, y_n)$ at our disposal. Therefore, we will confine ourselves to an empirical risk mitigation. For this purpose we determine step by step for each assignment function f of the class $\mathcal{F}$ the empirical training error when learning from a sample with scope n:

$$R_{\text{emp}}^n [f] = \frac{1}{n} \sum_{i=1}^{n} \frac{1}{2} |f(x_i) - y_i|$$

This creates a sequence of functions in class $\mathcal{F}$ with improved training error. The central question is whether this method can be used to determine pattern recognition with a minimum possible error deviation. Mathematically, the problem is whether the function sequence thus determined converges in the class $\mathcal{F}$ against a function with minimal error deviation.

In fact, it can be proven that such convergence or learning success is only guaranteed for small subclasses. An example is the Vapnik-Chervonenkis (VC) dimension, with which the capacity and size of such function classes can be determined [2]. With great probability, the risk is not greater than the empirical risk (plus a term that grows with the size of the function class).

The current successes of machine learning seem to confirm the thesis that it depends on the largest possible amount of data that is processed with ever-increasing computer power. However, the regularities detected then depend only on the probability distribution of the statistical data.

▶ **Definition**
Statistical Learning tries to derive a probabilistic model from a finite amount of data from results (e.g., random experiments) and observations (Fig. 11.1).

Statistical reasoning attempts, conversely, to derive properties of observed data from an assumed statistical model (Fig. 11.1).

Data correlations can provide indications of facts, but do not have to do so. Imagine a series of tests that show a favorable correlation between an administered chemical substance and the control of certain cancer tumors. Then there is pressure from the company concerned to go into production with an appropriate drug and skim off profits. But, affected patients may also see this as their last chance. In fact, we can only obtain a sustainable drug if we have understood the underlying causal mechanism of tumor growth, i.e. the natural laws of cell biology and biochemistry.

Already Newton was hardly interested in data correlations of the falling apples on the apple trees of his father's farm, but

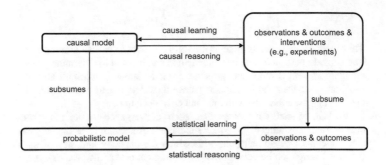

Fig. 11.1 Statistical and causal learning [3]

in the underlying mathematical causal law of gravitation, which made possible precise explanations and prognoses of the falling apples and celestial bodies, ultimately also the satellites and rocket technology based on them.

Statistical learning and reasoning from data is therefore not sufficient. Rather, we must recognize the causal relationships between causes and effects behind the measured data. These causal relationships depend on the laws of the respective application domain of our research methods, i.e. the laws of physics in Newton's example, the laws of biochemistry and cell growth in cancer research, etc. If it were otherwise, we could already solve the problems of this world with the methods of statistical learning and reasoning. In fact, some short-sighted contemporaries seem to believe this in the current hype of artificial intelligence.

▶   Statistical learning and reasoning without causal domain knowledge is blind—no matter how large the amount of data (big data) and the computing power!

In addition to the statistics of the data, additional laws and structural assumptions of the application domains are required, which are verified by experiments and interventions. Causal explanatory models (e.g., the planetary model or a tumor model) fulfil the law and structural assumptions of a theory (e.g. Newton's gravitational theory or the laws of cell biology):

▶ **Definition**
In causal reasoning, the properties of data and observations are derived from causal models, i.e. legal assumptions of causes and effects. Causal inference thus makes it possible to determine the effects of interventions or data changes (e.g., through experiments) (Fig. 11.1).

Causal learning, vice versa, tries to create a causal model from observations, measurement data, and interventions (e.g., experiments), which require additional laws and structural assumptions (Fig. 11.1).

A structural causal model consists of a system of structural allocations of causes to effects with possible noise variables. Causes and effects are described by random variables. Their functional assignments (under consideration of noise variables) are defined by equations, e.g., effect $X_j = f(X_i, N)$ in functional dependence on cause X_i and noise variable N. The network of causes and effects can be graphically represented by nodes and edges. Random variables of causes and effects correspond to nodes. Causal effects correspond to directional arrows: $X_i \rightarrow X_j$ means that cause X_i triggers effect X_j.

It can be proved that a causal model includes a clear probability distribution of the data (Fig. 11.1: "subsumed"), but not vice versa: For causal models (e.g., planetary model) additional laws (e.g., gravitational law) must be assumed [4]. In order to identify causal dependencies on events, the independence of the random variables representing them must be determined. Statistically speaking, the independence of the results x and y of two random variables (intuitively random experiments) X and Y can be expressed by the fact that their joint probability $p(x, y)$ is factorizable, i.e. $p(x, y) = p(x)p(y)$. This case is also called the Markov condition. On this basis, it is possible to introduce the calculus of a causal independence relation $\perp\!\!\!\perp$ [5]:

Background Information

Let $p(x)$ be the density of the probability distribution P_X of a random variable X:

X independent of $Y (X \perp\!\!\!\perp Y) :\Leftrightarrow p(x, y) = p(x)p(y)$

$$\text{for all values } x, y \text{ of } X, Y$$

$X_1, \ldots, X_d$ mutually independent $:\Leftrightarrow p(x_1, \ldots, x_d) = p(x_1) \cdot \ldots \cdot p(x_d)$

$$\text{for all values } x_1, \ldots, x_d \text{ of } X_1, \ldots, X_d$$

X independent of Y under condition $Z (X \perp\!\!\!\perp Y | Z) :\Leftrightarrow p(x, y|z) = p(x|z)p(y|z)$

$$\text{for all values } x, y, z \text{ of } X, Y, Z \text{ of } p(z) > 0.$$

Conditional independence relations fulfil the following rules:

$$X \perp\!\!\!\perp Y | Z \Rightarrow Y \perp\!\!\!\perp X | Z \qquad \text{(symmetry)}$$

$$X \perp\!\!\!\perp Y, W | Z \Rightarrow X \perp\!\!\!\perp Y | Z \qquad \text{(decomposition)}$$

$$X \perp\!\!\!\perp Y, W | Z \Rightarrow X \perp\!\!\!\perp Y | W, Z \qquad \text{(weak union)}$$

$$X \perp\!\!\!\perp Y | Z \text{ and } X \perp\!\!\!\perp W | Y, Z \Rightarrow X \perp\!\!\!\perp Y, W | Z \qquad \text{(contraction)}$$

$$X \perp\!\!\!\perp Y | W, Z \text{ and } X \perp\!\!\!\perp W | Y, Z \Rightarrow X \perp\!\!\!\perp Y, W | Z \qquad \text{(intersection)}$$

Example

Causal structural model with assignments and graphical representation [6]

$X_1 := f_1(X_3, N_1)$
$X_2 := f_2(X_1, N_2)$
$X_3 := f_3(N_3)$
$X_4 := f_4(X_2, X_3, N_4)$

N_1, N_2, N_3, N_4 independent noise variables

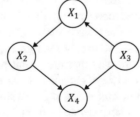

The independence of the random variable X_1, X_2, X_3, X_4 in the statistical distribution P_{X_1,X_2,X_3,X_4} can be represented by $X_2 \perp\!\!\!\perp X_3 | X_1$ and $X_1..X_4 | X_2$, X_3 or by the Markov factorization:

$$p(x_1, x_2, x_3, x_4) = p(x_3)p(x_1|x_3)p(x_2|x_1)p(x_4|x_2, x_3).$$

Objective of causal learning is therefore to discover the causal dependencies of causes and effects behind the distribution of measurement and observation data. The initial situation is a finite sample of a data collection: In Fig. 11.2, a joint probability (e.g., P_{X_1,X_2,X_3,X_4}) of independently and identically distributed (i. i. d.) random variables (e.g., X_1, X_2, X_3, X_4) is assumed. Through independence tests and experiments, causal models can be derived which are determined by independence relations, factorization, or causal laws. On the basis of such causal models, the dependencies of causes and effects can be graphically represented. This ensures the assignment (accountability) of causes and effects, which is necessary to clarify questions of responsibility.

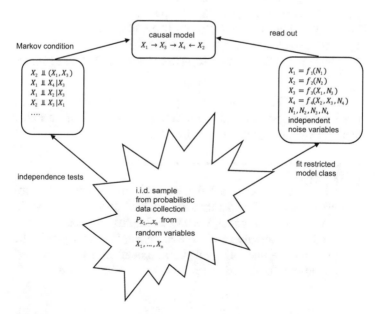

Fig. 11.2 From data mining to causal models [7]

The neural networks used practically in machine learning must have a large number of nodes. The number of possible causal models (with corresponding graphical representation) thus increases exponentially [8]:

d	Number of causal models with d nodes
1	1
2	3
3	25
4	543
5	29281
6	3781503
7	1138779265
8	783702329343
9	1213442454842881
10	4175098976430598143
11	31603459396418917607425
12	521939651343829405020504063
13	18676600744432035186664816926721
14	1439428141044398334941790719839535103
15	237725265553410354992180218286376719253505
16	83756670773733320187699303047996412223522313 8303

…

Because of this explosion of parameters, the complexity of practical applications leads to a dramatic challenge in machine learning which is often underestimated:

Example

In brain research, we are dealing with one of the most complex neuronal networks that has evolved during evolution. In the mathematical model, we take a simplifying vector z which encodes the activity of a large number of brain regions. The dynamics (i.e. the temporal development) of z is determined by a differential equation

$$\frac{d}{dt}z = F(z, u, \theta)$$

with F given function, u vector of the external stimulations and θ parameters of causal links [9].

But, brain activity z but cannot be observed directly. Functional resonance imaging fMRI determines only the consumption of nutrients (oxygen and glucose) to compensate for the increased energy demand supplied by blood flow (hemodynamic response). The growth is controlled by the blood-oxygen-level-dependent (BOLD) signal. Therefore, z in the dynamic causal model must be replaced by a state variable x in which the brain activity is taken into account with the thermodynamic response:

$$\frac{d}{dt}x = f(x, u, \theta).$$

For this purpose, the measured time series of the BOLD signal $y = \lambda(x)$ is connected with the state variable x.

In fact, in the human brain, we are dealing with a flood of data produced by 86 billion neurons. How in detail the causal interactions take place behind these data clouds remains a black box for the time being. Statistical Learning But even in the age of big data and growing computing power, statistical learning from measured data is not sufficient. More explanation of the causal interactions between the individual brain regions, i.e. causal learning, is a central challenge of brain research to achieve better medical diagnosis, psychological and legal accountability [10].

Example

Self-learning vehicles are a highly topical technical example of the growing complexity of neural networks. A simple automobile with various sensors (e.g., neighborhood, light, collision) and engine equipment can already generate complex behavior through a self-organizing neural network. If adjacent sensors are excited in a collision with an external object, then the neurons of a corresponding neural network connected to the sensors are also excited. This creates a wiring pattern in the neuronal network that represents the external object. In principle, this process is similar to the perception of an external object by an organism—only there much more complex.

If we now imagine that this automobile will be equipped with a "memory" (database) with which it can remember such dangerous collisions in order to avoid them in the future, then we can imagine how the automotive industry will be on the way in the future to build self-learning vehicles. They will differ significantly from conventional driver assistance systems with pre-programmed behavior under certain conditions. It will be neural learning, as we know it in the nature of more highly developed organisms.

But how many real accidents are necessary to train self-learning ("autonomous") vehicles? Who is responsible if autonomous vehicles are involved in accidents? Which ethical and legal challenges turn themselves in? In complex systems such as neural networks with millions of elements and billions of synaptic connections, the laws of statistical physics allow global statements to be made about the trend and convergence behavior of the entire system. However, the number of empirical parameters of the individual elements may be so large that no local causes can be identified. The neuronal network remains a "black box" for us. From an engineering standpoint, authors therefore speak of a "dark secret" at the heart of the AI of machine learning. Even the engineers who designed (the machine learning -based system) may struggle to isolate the reason for any single action [11].

Two different approaches in software engineering are conceivable:

1. Testing shows only (randomly) found errors, but not all other possible ones.
2. In order to avoid this, a formal verification of the neural network and its underlying causal processes would have to be performed.

The advantage of automatic proofs (Sect. 3.4) is to prove the correctness of a software as a mathematical theorem [12]. That's what proof assistants do [13, 14]. Therefore, the proposal is to create a formal meta-level on top of the neural network of machine learning, in order to realize proofs of correctness with a proof assistant automatically [15]. For this purpose, we imagine a self-learning automobile equipped with sensors and a connected neural network—quasi as the nervous system and brain of the system. The aim is to ensure that the behavior of the car follows the rules of the road traffic regulations. The road traffic regulations were formulated in 1968 in the Vienna Convention.

In a first step, the car will be like an aircraft with a black box to register the abundance of behavioral data. This mass of data should be implied by the traffic rules of the Vienna Convention. This logical implication realizes the desired control in order to rule out misconduct. At the meta-level, the implication is formalized to automate its proof through a proof assistant.

To this end, the legal system of the Vienna Convention would first have to be formalized. In a next step, the movement path, i.e. the causal trajectory of the vehicle, would have to be extracted from the mass of data in the black box. For this purpose, the causal learning can be applied. The causal trajectory can be represented graphically in a causal chain of causes and effects. This representation of the vehicle's trajectory would have to be represented on the meta-level in a formal language. This formal description would have to be implied by the formalized laws of the Vienna Convention. The formal proof of this implication is automated by the proof assistant and could be realized in a flash with today's computing power.

In summary follows: statistical machine learning with neural networks works, but we cannot understand and control the processes in the neural networks in detail. Today's machine learning techniques are mostly based only on statistical learning, but that is not enough for safety-critical systems. Machine Learning should therefore be combined with proof assistants and causal learning. In this case, correct behavior is guaranteed by metatheorems in a logical formalism.

This model of self-learning vehicles is reminiscent of the organization of learning in the human organism Behavior and reactions are also largely unconscious there. "Unconscious" means that we are not aware of the causal processes of the musculoskeletal system controlled by sensory and neural signals. This can be automated with algorithms of the statistical learning. But, this is not enough in critical situations: In order to achieve more safety through better control in the human organism, the mind must intervene with causal analysis and logical reasoning. Our goal is to ensure that this process is automated in machine learning by algorithms of causal learning and logical proof assistants.

Verification of software has already been a crucial step in the development of computer programs in software engineering [16]. After requirement engineering, design, and implementation of a program, different verification procedures have been applied in practice. A program is said to be correct ("certified") if it can be verified to follow a given specification which is derived from the design of the program. In Sect. 3.4, we already considered proof assistants which prove the correctness of a computer program in a consistent formalism like a constructive proof in mathematics (e.g., Coq, Agda, MinLog). Obviously, proof assistants are the best formal verification of correctness for certified programs.

But, in practice of industry and business, proof assistants seem to be too ambitious because of the increasing complexity of AI-software. Therefore, industrial production is often content with ad hoc tests or empirical test procedures which try to find statistical correlations and patterns of failures. Empirical testing [17] lays directly on the analysis of program executions. It collects information from executing the program either after actively soliciting some executions, or passively during operation and try to abstract from these some relevant properties of data or of behavior. On this basis, it is decided whether the system conforms to the expected behavior.

Model-based testing does not only rely in empirical data mining [18]. It uses a model of the implementation of a technical system that is based on its design. From this implementation model, test input is automatically generated and executed by a test tool. The output of the system is automatically compared to the output specified by the model of the system. In this step, the conformance of the implementation with the specification of the system is tested. If the system passes all the generated tests, then the system is considered to be correct.

A test tool architecture of model-based testing starts with a test engine which implements the test generation procedure: It steps through the specification of the model and computes the sets of allowed input and output actions. If an output action is observed, then the test engine evaluates whether this output is allowed by the specification of the model. In this step, the conformance of implementation and specification is verified [19].

If some output is observed that is not allowed according to the specification, then the test is terminated with the verdict fail. As long as the verdict fail is not given, the test terminates with the verdict pass. Now, we can define the input-output conformance of the implementation and specification of a technical system.

▶ An input enabled implementation conforms to a specification if after every trace of input actions and output actions of the specification, the output actions allowed by the implementation form a subset of the output actions allowed by the specification.

An even more rigorous verification of a program would be realized by a proof assistant (e.g., Coq, Agda, Minlog, Isabelle). In this case, the models of implementation and specification are transformed into a constructive formalism (e.g., CiC in Coq) and the conformance of implementation and specification is proved in this formalism according to the exact standards of formal theorem proving in mathematical logic. In ideal cases, these proofs are even automated.

A challenge is the verification of machine learning with neural networks and learning algorithms. The increasing complexity of neural networks with an exploding number of parameters generate black boxes which seem only to be trainable by Big Data and testable by ad hoc and empirical procedures. But, in Sect. 10.2, we explained why neural networks are mathematical equivalent to automata and Turing machines with respect to certain cognitive abilities [20]. Therefore, the technical implementation of a neural network (e.g., an electrical circuit) and its specification resp. its expected or desired behavior can also be represented by automata or machines. At least in principle, they could be transformed into the formalism of a proof assistant, e.g. the Calculus of inductive Constructions (CiC) in Coq.

Thus, in principle, the conformance of technical implementation and specification could mathematically be proved in the CiC formalism. In that sense, the correctness of neural networks could be rigorously proved in Coq. Even analog neural networks (with real weights) could be implemented into CiC extended by higher inductively defined structures to verify their correctness in Coq. The challenge is, of course, the complexity of practically applied neural networks in machine learning.

Figure 11.3 illustrates a scaling of software testing procedures from ad hoc testing by random samples with low degrees of accuracy, empirical testing (anti-model-based testing) with more reliable results up to model-based-testing and finally proof assistants with the highest degrees of accuracy, but until now less applicability because of increasing complexity of AI-software.

Therefore, in practice, we must also consider the costs of testing. Formal proofs of complex software need an immense amount of time and man-power. On the other side, it is risky to rely in ad hoc testing and empirical testing only in safety-critical systems. For certification of AI-programs, we must aim at increasing accuracy, security, and trust in software in spite of increasing complexity of civil and industrial applications, but

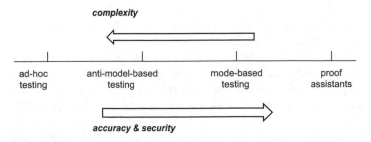

Fig. 11.3 Degrees of certification in software testing

with respect to the costs of testing (e.g., utility functions for trade-off time of delivery vs. market value, cost/effectiveness ratio of availability). There is no free lunch for the demands of safety and security. Responsible AI must find fair and sustainable degrees of certification.

11.2 Deciding Under Incomplete Information

In complex markets, people do not behave according to the axiomatically defined rational expectations of a "representative agent" (homo oeconomicus), but decide and act with incomplete knowledge, emotions, and reactions (e.g., herd behavior). Therefore, the American Nobel Prize winner Herbert A. Simon (1916–2001) spoke of bounded rationality [21]. What this means is that we should be satisfied with solutions that are satisfiable for the time being and not strive for perfect solutions in the face of complex masses of data.

But, do decisions under bounded rationality and information remain, in principle, closed to an algorithmic determination? In this context, it is remarkable that an AI software can beat human champions in poker [22]. Poker is spectacular for several reasons. Unlike board games like chess and Go, poker is an example of decisions made under incomplete information. Of exactly this type are everyday decisions that take place under incomplete information at, e.g., negotiations between companies, legal cases, military decisions, medical planning, cybersecurity etc. Board games such as chess and Go, on the other hand, involve decisions in which each player has a complete overview of the entire game situation at all times. In poker, you always suspect emotions and feelings in the game to deceive the opponent with poker face due to incomplete information. But, until machines will be able to understand or even realize human emotions— even AI experts thought—many more years would pass, if it should succeed at all. In fact, Poker Libratus avoids the problem of emotions and beats people with sheer computer power plus sophisticated mathematics.

▶   Artificial intelligence does not have to imitate human
     intuition and emotion in order to be able to beat a
     person when making decisions based on incomplete
     information.

At this point, it becomes clear that technically successful AI is
above all an engineering science that wants to solve problems
efficiently. It is not about modelling, simulating or even replacing human intelligence. Even in the past, successful engineering
solutions were not designed to imitate nature: As long as people
tried to imitate the flapping wings of birds, flying went wrong.
It was only when engineers began to think about the basic laws
of aerodynamics that they found solutions for moving aircraft
weighing several tons—solutions that were not found by nature
in evolution. Engineering AI must be distinguished from brain
research and neuromedicine, which want to model and understand the human organism—as it developed in natural evolution.

A game process or a negotiation situation is represented
graphically by a game tree. A game situation corresponds to a
node, from which, according to the rules of the game, there are
finally many moves, which are represented by corresponding
branches in the game tree. These branches end again with nodes
(game situations), from which new possible branches (game
moves) spring again. This is how a complex game tree unfolds.

In a first approach, an effective algorithm selects the weak-
nesses of a previous game in the corresponding game tree and
tries to minimize them in subsequent games (game trees). The
system does not play against itself ten, a hundred or a thousand times, but millions of times due to the enormous computing power of a supercomputer. At approx. 10^{126} even the fastest
supercomputers wouldn't be able to do that in any realistic time.
Mathematics is now being used: with theorems of mathematical probability and game theory, it can be proven that in certain
game situations there is no chance of success for subsequent
games. Therefore, they can be neglected in order to reduce computing time.

With this in mind, we can distinguish between two algorithms
in Poker Libratus [23]: Counterfactual Regret Minimization (CFR)

is an iterative algorithm to solve zero-sum games with incomplete information. Regret-Based Pruning (RBP) is an improvement that allows the developmental branches of less successful actions in the game tree to be "pruned" temporarily to speed up the CFR algorithm. Based on a theorem by N. Brown and T. Sandholm (2016), in zero-sum games, RBP truncates any action that is not part of a best response of a Nash equilibrium. A Nash equilibrium is a game constellation in which no player can improve his score through a one-sided strategy.

Therefore, in games with incomplete information, one tries to find a Nash equilibrium. In 2-person zero-sum games with less than approx. 10^8 possible game constellations (nodes in the game tree), a Nash equilibrium can be found exactly by a linear algorithm (computer program). For larger games one uses iterative algorithms (e.g., CFR), which converge to a Nash equilibrium as limit value.

After each game, CFR calculates the regret of an action at each decision point of a game tree, minimizing the degree of regret and improving the game strategy: "Contrafactual" means "What could have been done better?" If an action is associated with negative regret, RBP skips that action for the minimum number of iterations necessary until the associated regret becomes positive in CFR. The skipped iterations are done in a single iteration as soon as the pruning is finished. This leads to a reduction in computing time and storage space, which can be realized by today's physical machines.

▶ AI software, which generates extensive empirical knowledge faster and more effectively on the basis of large amounts of data with increasing computer power, becomes indispensable for human decision-making when information is incomplete. In negotiating situations that are confusing for people, this type of AI software will be able to examine the possibilities of winning strategies and propose favorable decisions. This will be just as important in complex corporate negotiations as it will be in supporting difficult

legal decisions. However, the increasing efficiency
of this software also challenges people's judgement:
The possibilities and limitations of this software must
be precisely determined in foundational research in
order not to fail by blindly relying on misunderstood
algorithms.

11.3 How Secure Are Human Institutions?

The exponential growth of computing power will accelerate
the algorithmization of society. Algorithms will increasingly
replace institutions and create decentralized service and supply
structures. The database technology blockchain offers an entry
scenario for this new digital world [24]. It is a kind of decentralized accounting, which, e.g., replaces banks for the mediation
of money transactions between customers by algorithms. This
decentralized agency was invented after the global financial crisis of 2008, which was largely caused by human error in national
and international central banks.

Blockchain can be presented as accounting via a continuous
decentralized database [25]. The bookkeeping is not centrally
stored, but is stored as a copy on every computer of the partici-
pating actors. On each "page" (block) of the accounts, transactions between the actors and security codes are recorded until
they are "full" and a new page has to be "opened". Formally,
it is an expandable list of data records (blocks) that are linked
with cryptographic procedures (Fig. 11.4). Each block contains
a cryptographically secure hash of the previous block, a times-
tamp, and transaction data. New blocks are created by a consensus procedure (e.g. Proof-of-Work algorithm).

▶   By the accounting system "blockchain", digital goods
      or values (currencies, contracts, etc.) can be repro-
      duced at will: "Everything is a copy". After the Internet
      of Things (IoT = Internet of Things, cf. Chap. 9), the
      Internet of Values (IoV = Internet of Values) is thus
      announced.

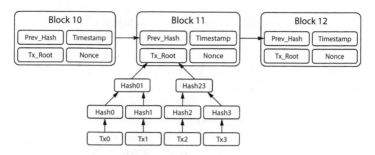

Fig. 11.4 Block chain with hash coding

Due to the sequential storage of data in blockchains, one-sided changes are immediately recognizable. Each actor involved would recognize changes in his copy of the blockchain, since for this the blocks linked into each other would have to be "unpacked". In addition, the high computing capacity of the entire network in "block mining" makes blockchains virtually forgery-proof. A decentralized crypto currency works in the following steps [26]:

1. New transactions are signed and sent to all nodes of the actors.
2. Each node (actor) collects new transactions in a block.
3. Each node (actor) searches for the so-called nonce (random value) that validates its block.
4. If a node (actor) finds a valid block, it sends the block to all other nodes (actors).
5. The nodes (actors) only accept the block if it is valid according to the rules:
   a) The hash value of the block must correspond to the current difficulty level.
   b) All transactions must be signed correctly.
   c) The transactions must be covered according to the previous blocks (no duplicate output).
   d) New issues and transaction fees must comply with accepted rules.

6. The nodes (actors) express their acceptance of the block by adopting its hash value into their new blocks.

Creating a new valid block (mining) corresponds to solving a cryptographic task (proof-of-work). The difficulty of the task is regulated in the network in such a way that a new block is created every 10 min on average. The probability of successful mining is proportional to the computing power used. For this purpose, the degree of difficulty of the mining must be constantly adapted to the current computing power of the network. The proof-of-work algorithm runs in the following steps. The threshold used is inversely proportional to the mining difficulty [27]:

1. initialize block, calculate root hash from transactions
2. calculate hash value: h = SHA256(SHA256(block header))
3. If h >= threshold, change block header and return to step 2
4. else (h < threshold): valid block found, stop calculation and publish block.

▶ Definition The mining difficulty level is a measure of how difficult it is to find a hash value less than a given target. The Bitcoin network has a global block difficulty:

difficulty = difficulty_1_target/current target
(target is a number with 256 bit.)

The difficulty level is adjusted every 2016 blocks—appropriate to the time to find them.

With the desired rate of one block per 10 min, 2016 blocks require exactly two weeks.

time (2016 Blocks) >2 weeks ⇒   level of difficulty must be
                                 diminished
time (2016 Blocks) <2 weeks ⇒   level of difficulty must be
                                 heightened

▶ Definition The nonce value is a 32-bit field whose value is set so that the hash (key code) of the block in question begins with a sequence of zeros. The rest of the field (with a defined meaning) must not be changed:

- Since it is considered impracticable to predict the bit combination of the correct hash, many different hash values must be attempted:
- The hash value is calculated for each nonce value until its hash with the required number of zeros is found.
- The number of zero bits required is determined by the degree of mining difficulty.
- The resulting hash must have a value less than the current mining difficulty level.
- Since this iterative calculation requires time and resources, the presentation of the block with the correct nonce value proves the required workload (proof-of-work).

The transactions contained in the new block are initially confirmed only by the participant who created the block. They have only limited credibility. However, if the block has also been accepted as valid by the other participants, they will enter its hash value in their new blocks to be created. If the majority of participants consider the block to be valid, the chain will continue to grow fastest from this block. If they do not consider it valid, the chain will continue to grow from the previous block. Therefore, the blocks form a tree.

Only the chain longest in the first block (root) of the tree is considered valid. Thus, this form of accounting automatically consists of those blocks that have been accepted as valid by the majority. This first block, which is used to start a crypto currency, is called the Genesis block. It is the only block that does not contain a hash value of a predecessor.

The Bitcoin network is based on a decentralized database (blockchain) managed jointly by the participants using Bitcoin software, in which all transactions are listed. Instead of confidants and institutions (e.g., banks, state currency control, central banks), computationally complex and practically forgery-proof algorithms are used (e.g. proof-of-work algorithm). Proof of ownership of Bitcoin can be stored in a personal digital wallet. Bitcoin's conversion rate to other means of payment is determined by supply and demand. This can trigger speculative bubbles, which is currently still a problem for the general acceptance of Bitcoin.

▶   Block chain will be an entry-level technology for a decentralized digital world in which people as customers and citizens realize their transactions and communications directly and without intermediary institutions.

The perspective of this technology is by no means limited to banks and monetary transactions. Future developments are also conceivable, in which other service facilities and state institutions are replaced by algorithms. What at first glance appears to be very grassroots democracy, turns out to be anything but democratic on closer analysis. The basic idea of democracy is that regardless of their position and arrival, everyone has only one vote: One man—one vote! In fact, the power of influence at Bitcoin depends on the computing power with which a customer asserts himself in the realization of a new block: the greater the available computing power, the greater the probability and confidence that someone can solve the cryptographic task and thus guarantee security (proof-of-work).

With growing blockchain, these tasks become more and more complex and computationally intensive. But computing intensity is also energy-intensive. The fact that computation-intensive algorithms consume enormous amounts of energy is hardly considered. In November 2017, Bitcoin's computing network consumed as much kilowatts per hour as the entire country of Denmark. Therefore, countries with cheap energy and cooling for hot supercomputers can produce most Bitcoins (e.g., China). Unless countermeasures are taken and improvements made, such infrastructures in no way promise the salvation of a direct democracy, but rising energy problems (and thus growing environmental problems). In the end, digitization depends on the overall balance of better infrastructure, less energy consumption, a better environment, and more democracy.

References

1. Peters J, Janzing D, Schölkopf B (2017) Elements of causal inference. Foundations and learning algorithms. The MIT Press, Cambridge, p 6 f.
2. Vapnik VN (1998) Statistical learning theory. Wiley, New York
3. Peters et al (Note 1), 6, after Figure 1.1
4. Mooij JM, Janzing D, Schölkopf B (2013) From ordinary differential equations to structural causal models: the deterministic case. Proceedings of the 29th Annual Conference on Uncertainty in artificiaL Intelligeence (UAI), pp 440–448
5. Pearl J (2009) Causality: models, reasoning, and inference. Cambridge University Press, Cambridge
6. Peters et al (Note 1), p 84, after Figure 6.1
7. Peters et al (Note 1), p 144, after Figure 7.1
8. Foundation Inc OEIS (2017) The on-line encyclopedia of integer sequences. http://oeis.org/A003024.2017. Accessed 24 Nov 2018
9. Friston K, Harrison I, Penny W (2003) Dynamic causal modelling. Neuroimage 19:1273–1302
10. Lohmann G, Erfurth K, Müller K, Turner R (2012) Critical comments on dynamic causal modelling. Neuroimage 59:2322–2329
11. Knight W (2017) The dark secret at the heart of AI. MIT Technol Rev 11:1–22
12. Mainzer K (2018) The digital and the real world. Computational foundations of mathematics, science, technology, and philosophy. World Scientific, Singapore (Chapter 7)
13. Schwichtenberg H (2006) Minlog. In: Wiedijk F (ed) The seventeen provers of the world. Lecture notes in artificial intelligence, vol 3600. Springer, Berlin, pp 151–157
14. Nipkow T, Paulson LC, Wenzel M (2002) Isabelle/HOL. A proof assistant for higher-order logic. Springer, Heidelberg
15. Mainzer K (2018) How predictable is our world. Challenges for mathematics, computer science, and philosophy in the age of digitalization. Springer, Wiesbaden
16. Bourque P, Dupuis R (2004) SWEBOK. Guide to the software engineering body of knowledge. IEEE Computer Society, Los Alamitos
17. Bertolino A (2007) Software testing research: achievements, challenges, dreams. In: Future of Software Engineering (FOSE'07) 0-7695-2829-5/07 IEEE

18. Tretmans J (1996) Test generation with inputs, outputs, and repetitive quiescence. Softw Concepts Tools 17(3):103–120
19. Tretmans J, Brinksma E (2003) TorX: Automated model-based testing. In: Hartman A, Dussa-Zieger K (eds) Proceedings of the First European Conference on Model-Driven Software Engineering 2003
20. Mainzer K (2018) The digital and the real world. Computational foundations of mathematics, science, technology, and philosophy. World Scientific Publisher, Singapore (chapter 12)
21. Simon H (1947) Administrative behavior: a study of decision-making processes in administrative organizations. Macmillan, New York
22. Bowling M, Burch N, Johanson M, Tammelin O (2015) Heads-up holdem poker is solved. Science 347(6218):145–149
23. Brown N, Sandholm T (2017) Reduced space and faster convergence in imperfect information games via pruning. International Conference on Machine Learning (ICML)
24. Economist Staff (2015) Blockchains: The great chain of being sure about things. The Economist October 31st
25. Narayanan A, Bonneau J, Felten E, Miller A, Goldfeder S (2016) Bitcoin and cryptocurrency technologies. A comprehensive introduction. Princeton University Press, Princeton
26. crypto currency. Wikipedia. https://de.wikipedia.org/wiki/Kryptow%C3%A4hrung. Accessed 24 Nov 2018
27. Wikipedia (2017) Bitcoin. https://de.wikipedia.org/wiki/Bitcoin. Accessed 24 Nov 2018

Artificial Intelligence and Responsibility

Artificial intelligence (AI) is an international future topic in research and technology, economy, and society. But research and technical innovation at AI are not enough. AI technology will dramatically change the way we live and work. The global competition of social systems (e.g., Chinese state monopolism, US-American IT giants, European market economy with individual freedom rights) will depend decisively on how we position our European value system in the AI world.

12.1 Social Score and New Silk Road

Supercomputers and machine learning are already seen as the key to world domination in countries like China. Even the use of nuclear weapons is of secondary importance, as the computation of the necessary data volumes and strategic planning depends on powerful computers. In a nutshell: The nuclear age was yesterday; today and tomorrow we are talking about digitization and artificial intelligence. Carl Friedrich von Weizsäcker had emphasized the responsibility of the scientist in the atomic age [1]. This question intensified at that time against the background of the Cold War between the West and the Eastern bloc. Today we are talking about responsibility in the age of digitalization and artificial intelligence. Before we move on to the next Sect. 12.2, the changed global political situation will first be examined.

© Springer-Verlag GmbH Germany, part of Springer Nature 2020   267
K. Mainzer, *Artificial intelligence – When do machines take over?*, Technik im Fokus,
https://doi.org/10.1007/978-3-662-59717-0_12

It became clear during the Second World War that mathematical methods, together with computer power, are decisive for military purposes. The history of the British and Polish mathematicians, logicians, and cryptologists who succeeded in decrypting the German encryption machine Enigma is well known [2]. The decryption contributed considerably to the shortening of the war during the air battle for England, but especially during the submarine war. The encryption machine Enigma (Greek: secret) was not only used to encrypt military messages, but also by other government agencies of National Socialist Germany. So it already affected what we now call infrastructure. From 1939, Alan Turing worked as a cryptanalyst at Bletchley Park, the headquarter of the British code breakers. The processes in a Turing machine had inspired Turing to develop his decoding process.

Strategic planning requires the use of powerful forecasting tools in politics, business, and the military. This is where machine learning comes in.

In 1965, the physicist Wilhelm Fucks published his then bestseller "Formeln zur Macht" ("Formulas to Power") [3]. Remarkable predictions have been derived from simple growth equations: China would rise to a superpower in the foreseeable future and displace the USA by far. This prediction was not to be expected easily at the height of the Cold War, when the USA and the Soviet Union, as the two leading world powers, faced each other militarily upgraded. Crisis points are predictable after Fucks as metastable equilibria. Alliances provide calculable advantages. In 1965, population growth, steel, and energy production were the most important selected indicators—certainly a one-sided selection from today's perspective. But, we are talking about mathematical models with if-then-statements: If certain assumptions apply, then the logically-mathematically derived events follow with a certain probability. The book also showed the post-war generation how completely illusory the ideology of world conquest pursued by the Germans in World War II was due to the size of the country—quite apart from the terrible moral crimes associated with it. However, the book title "Formeln zur Macht" (Formulas to Power) was dazzling, reminding us of Nietzsche's "Will to Power". The mathematical methods of the book seemed to announce the counterprogram: Politicians can "want" a lot of things. In the atomic age of that time, it was Einstein's formula $E = mc^2$ that made the world tremble. Those who do not understand the language of mathematics cannot understand this world [4].

Prognosis with Big Data seems at first to be harmless and even useful: Healthcare and insurance companies collect large amounts of data on sporty exercise, nutrition, and alcohol consumption to predict likely treatment costs or even time of death. In the USA, forecasting systems are used to make more or less favorable social forecasts for prisoners in order to release them on probation. For this purpose, huge databases with the data on probation of released prisoners will be created, in order to extract statistical patterns and correlations about socially favorable behavior using big data algorithms.

What is to be realized in China under the title "Social Score" from 2020, however, is once again of a completely different dimension [5]: All citizens are to be captured in a total evaluation of their social behavior. Now in Germany we already know the infamous "points in Flensburg" with which traffic sins are punished. Writings and achievements of university teachers and researchers are internationally evaluated with so-called impact factors, which are used for appointments. Many citizens in the Western countries get excited about data on a health card. But, social score à la Chinoise evaluates everything and everyone in a public scoring system that can decide on loans and housing rental as well as professional promotions or state benefits. In the end, one could be a more or less socially valuable person, whose life in old age might enjoy more or less great care support.

The National Security Agency (NSA) of the USA naturally does not want to be inferior and tries to calculate the behavior of a politician in a crisis with millions of equations. The individual equations are usually by no means complicated, but linear and refer to the individual measured behavior parameters. But, it is the mass of these data that makes supercomputers calculate for hours. This is where the Chinese government comes in and wants its supercomputer Tianhe to be completed by 2020, in order to double the computing time of the currently fastest computer Summit from the USA with 1.5 trillion computing steps per second to 3 trillion [5].

With this technology, it should then be possible to predict the behavior of individual citizens on the basis of their social score in just a few seconds. Efficiency is to be further increased by reducing the necessary linear equations to less than one million [6]. The ambition of political leadership, the military and intelligence services is not only directed towards domestic policy, but also towards foreign policy: in the end, one wants to recognize and predict future crises in good time in order to control and stabilize the state at home and abroad. Behind this are by no means the well-known horror ideas of a Big Brother, who strives for dark exploitation strategies of the masses in favor of a few. Rather, it is about a highly developed welfare state which, however, is capable of avoiding the destabilization and crisis susceptibility of Western democracies [7].

▶   This kind of "world revolution" is by no means to be
    introduced in an act of violence. Rather, it is assumed
    in China's strategy that the citizens of Western
    democracies will recognize and take over the benefits
    of stable and efficient technocracies in the final state
    of global competition.

In fact, it is assumed here, citizens are not "guided" by a political leadership in the traditional sense either. Rather, such a society controls itself on the basis of the "objectively collected, algorithmically perfectly calculated, and societally accepted" social score. After all, it is not a "party" of error-prone people who steers and controls, but a network of intelligent algorithms. What is already being announced today in autonomous car driving, where millions of road deaths are avoided through automation, will then continue on a grand political scale.

What may shock from a Western perspective and is (still?) rejected as a technocratic fantasy of omnipotence falls on a culturally different sounding board in Asia. In a society shaped by centuries of Confucianism, it is by no means alien that every member of society occupies a rank, i.e. a social score, according to generally accepted virtues according to merit and misconduct. Confucian virtue ethics by no means places individual liberty rights at the top of the hierarchy of values, but rather the common good, which all must serve according to their merits [8–10]. This is considered to be reasonable and has guaranteed China's stability for centuries, in contrast to its neighboring countries. Machine learning, big data, and predictability are seen as opportunities for modern technology to continue in this tradition.

The necessary forecasting systems are not only a good business for Chinese software houses. The large American Internet companies such as Google, Facebook, and Amazon also sense economically lucrative application possibilities for health systems, government authorities, and companies in their countries. Silicon Valley is still being considered a cadre center for worldwide artificial intelligence. But it could turn out, that in the end China (as Wilhelm Fucks 1965 predicted on another basis) will win the race: The Social Core Program is implemented as a strategic overall project of this huge country in a tight schedule with a huge capital expenditure to which all entrepreneurial and technological interests have to submit. From the Chinese point of view, the increasing inability of Western democracies is noticed to implement large structural projects. Therefore, Europe must show that it has a global IT and AI strategy of its own and does not fall into local parties and country squabbles. Individual European IT and AI trendsetters such as, the small Estonia as a lighthouse, are not enough.

▶   The key question in the global competition of artificial intelligence is whether an AI technocracy with (e.g., Chinese) state capitalism and Confucian ethics will prevail against Western market economy and democracy, in which AI systems are understood as a service of individual freedom rights.

The first system competition took place during the Cold War, when market-economy democracies competed against communist centralized administrative economies. It was about military supremacy to expand the respective political and economic system. In the final phase of the Eastern bloc, cybernetic models and computer technology were already being used massively in highly developed countries such Eastern Germany as the former DDR. But, the lack of efficiency of the economic system combined with the restriction of its citizens eventually leads to implosion. At the same time, a competition for locational advantages emerged in the Western democracies and market economies: Who has the better tax, wage, social, and educational system to offer better opportunities for innovation and investment? This competition between Western countries continues.

The local disputes are superimposed by the new global competition of Western democracies with a technocratic state capitalism, which can be found not only in China, but also to some extent in Russia and smaller Asian states such as Vietnam. In contrast to the first global competition, countries like China have now incorporated market-based elements such as private enterprise and free pricing. Among the world's 500 largest companies in 2018 are 103 companies from China, although the state has a majority stake in 73 companies. The banking sector is also largely under state control. However, while Western democracies are dependent on election periods, changes of governments, and the associated crises and destabilizations, countries such as China can implement major innovations and infrastructure projects (such as artificial intelligence) in the long term.

The rise of China was initially achieved with imports of cheap consumer goods, which led to job losses in the West. On the other hand, China opened up new sales markets for German key technologies such as automotive, motor industry, and electrical engineering. China is now Germany's most important trading partner. Foreign trade surpluses made China a rich country, which in a next step could switch to purchasing key industries (e.g., robotics, electromobility, biotechnology) in western countries and thus acquire the corresponding know-how. This includes a strategy project on a global scale that has once again underscored China's claim to be "the Middle Kingdom" for all other centuries—the Silk Road.

In the New Silk Road project, old trade routes between China and Europe are to be developed into infrastructures with energy, communication, and information networks, railway lines, ports, and roads [11]. China is investing in these infrastructure networks, which the respective riparian states would not be capable of. This creates progress for these countries, but also political dependencies. Infrastructures of this scale can only be controlled with the support of digitalization and artificial intelligence. Today, freight transport and logistics are already supported by robotics, sensor technology, and satellite communication. Trade and capital flows of the New Silk Road will continue to promote the transfer of ideas, innovation, and culture—now by means of information technology and artificial intelligence.

In order to compete with China, Europe must first insist on opening the Chinese market to its own products. The joint plea for free world trade is good, but must be fair in both directions. China itself officially speaks of a win-win situation. Only in this way, the New Silk Road will be able to secure mutual prosperity.

12.2 Artificial Intelligence and Global Competition of Value Systems

The social score of the Chinese AI technocracy presupposes a total data collection of all citizens. In Western democracies, the individual's rights to freedom are at the center of the value system. Its philosophical roots are declared as the autonomy of the person. In AI research, people are already talking about "autonomous" driving. In fact, it is still often just pre-programmed driver assistance systems. Also in the case of machine learning, the learning algorithms are given by the programmer, although the system can decide and learn from itself in given situations.

Another example are "moral" war robots (moral soldier) that the US military develops. The background is the experience with soldiers who were social citizens in civilian life, but who brutalized and started terrible war crimes after traumatic war experiences in the war (e.g., My Lai Massacre in the Vietnam War 1968). Similar to driving a car, the technical consideration is that human deficits can be avoided during emotional stress by using reliable AI systems that adhere to the rules in all cases—whether in traffic law or martial law.

However, a technical system that adheres to pre-programmed moral rules is not itself "moral". Even in the case of learning algorithms, we are at best dealing with AI systems that can be compared with trained animals or toddlers. But, autonomy in the sense of political freedom rights means a higher level:

▶ Autonomous is the man who is self-determined in every respect and capable of self-legislation.

According to Kant, an act is morally justified only if it can form the basis of general legislation [12]. This is the core idea of his categorical imperative: The rule (Kant: "maxim") of my action must be generalizable. My right ends where the right of the other begins. If I enter into the free space of the other, this action cannot be generalized: it would inevitably entail the war of all against all. Therefore, in principle, the maxim of my action should be the basis of a general law, which, e.g., a parliament decides on. In this sense, autonomy means the ability to "legislate oneself".

Technically, it cannot be ruled out that one day an artificial intelligence will also be capable of "self-legislation", i.e. it gives itself its laws as programs: it programs itself! In the separation of powers in Western democracies, legislation is the right of parliament (legislative power) elected by the population of a country in free elections. In Sect. 10.3, there was already talked about superintelligence, which has a better overview of global situations than the individual human being. In this case, however, we would have given up our autonomy. AI would no longer be just a service system; in Hegel's words, AI would no longer be a "servant" but the "master".

At this point, at the latest, the question of responsibility in the age of digitalization and artificial intelligence arises fundamentally. The concept of responsibility has a long legal and philosophical tradition [13].

▶ Responsibility is generally understood as the duty of an acting person (or group of persons) towards another person (or group of persons) on the basis of a claim asserted by an authority (e.g., court, state, society).

First distinguishing criteria are causal responsibility with regard to causation (e.g., programmer's programming error), role responsibility with regard to a task (e.g., teacher for his school class), skill responsibility with regard to fulfillability (e.g., a physician in an accident), and liability responsibility that may differ from causation (e.g., "parents are liable for their children") [14]. The determination of causal responsibility is not normative, but it is based on empirical findings. That is the central problem with the black boxes in neural networks that were mentioned in Sect. 11.1.

In questions of liability, legal entities (e.g., companies) are treated as responsible subjects. Criminal responsibility for institutions does not exist, however, under German law (in contrast to US-American law). At least morally, however, companies are also ascribed responsibility. This is referred to as corporate governance and corporate social responsibility (CSR).

In legal terms, responsibility is understood as the duty of a person to be accountable for his/her decisions and actions in accordance with specified regulations (accountability). Formally, therefore, law does not refer to moral or religious responsibility (in the of conscience), but ("positivist") to the violation of laws that is established by a court. Thus, responsibility in the legal sense is always tied to empirical facts. Therefore, the demand for more explainability of causal processes in machine learning is of fundamental importance for the clarification of legal responsibilities (see Sect. 11.1).

From a legal point of view, a distinction is made between accountability in the following aspects [15]:

a) Action responsibility is the term used to describe accountability with regard to the way in which tasks are performed.
b) Profit responsibility is the term used to describe accountability with regard to the achievement of objectives.
c) Leadership responsibility is the term used to describe accountability with regard to the management tasks performed, including the associated external responsibility.

In law, responsibility refers not only to persons, but also to property (e.g., computers) and to requirements of an owner, trustee, or lessee. With the increasing degree of autonomy of intelligent systems (cf. working definition of AI in chap. 1), the question arises to what extent robots can still be treated as material goods or whether we already have to legally consider intermediate areas between material goods and persons. Animal law shows how inappropriate the traditional distinction between thing and person is when we take modern findings of evolutionary biology and cognitive psychology as a basis: animals are living beings capable of suffering and not "things", but on the other hand not yet "persons" [16, 17].

Artificial intelligence is undoubtedly subject to the principle of responsibility: it is the human being who should determine how it is used. However, specialization and the growing complexity of technical, social, and ecological contexts lead to a diffusion of responsibility: the individual is increasingly dependent on the information and assessments of other experts. As a consequence, there is a need for the institutional attribution of responsibility by legal or contractual provisions, e.g., in liability law, and/or the attribution of responsibility to collective actors such as companies and associations. However, the diffusion of responsibility also favors clear violations of the law and misuse of technology, which lead to public outrage and uncertainty. However, security and trust in technology are prerequisites for the future viability of a country.

With regard to complex AI systems and AI infrastructures, the concept of responsibility needs to be expanded. In addition to individual responsibility, collective and cooperative responsibility must also be analyzed in terms of systems theory. Responsibility should also be ascribable to those who are responsible for the design of AI systems (e.g., Industry 4.0), the development of interfaces, platforms, and the use of infrastructure. In these cases, the degrees of influence are to be measured.

Responsibility for the future requires the early identification and assessment of risks. In the debate on responsibility for the future, Hans Jonas in particular emphasized the need to refrain from actions that could pose an existential threat to the environment or future generations [18]. This is especially true for artificial intelligence. But, nobody can predict future innovations like the astronomer the position of a planet. For this reason, instead of assessing the consequences of technology, we have raised the demand for technology design in Sect. 10.4. It must be linked to early education and training.

▶ **Important** To this end, we demand that the topic
of ethics and responsibility should be reflected in all
studies and curricula of a university dealing with AI
topics in suitable teaching formats. It is necessary to
give opportunities to deal with the ethical questions
of the AI responsibility to all students in all faculties
dealing with AI topics. The question of ethical responsibility must already be integrated into the studies,
just as ethical and social questions should later be
taken into account in research and technology development. This is the only way to achieve sustainable
technology design.

The scope and content of the ethical study formats
in AI should be agreed with the program coordina-
tors. In addition to general principles, the contents
must be linked to the concrete ethical questions of
the special AI degree programs in computer science,
engineering and natural sciences, medicine and
economics.

Ethics should therefore not be misunderstood as a brake on innovation. Rather, raising awareness of ethics and responsibility
promote innovation advantages such as greater legal certainty
and social acceptance of AI research in society. The focus is on
the international challenge of how AI systems should be understood as a service provided by democratic societies that wish to
continue to invoke their individual freedoms and human rights.
Internationally, the combination of AI and democracy should
strengthen the locational advantage of a country: A country
should not only be strong in AI innovation, but should also take
account of social responsibility issues.

▶ Europe must not only be a leader in AI innovation (e.g., at the interface of machine learning and
economy in industry 4.0), but to create an attractive
societal environment associated with it. Even in the
age of digitalization and artificial intelligence, the

protection of individual freedom rights and secure social systems in a market economy remain high goods that should be recognized and valued by all people worldwide.

References

1. von Weizsäcker CF (1978) The responsibility of science in the atomic age, 6th edn. Vandenhoeck & Ruprecht, Göttingen
2. Hinsley FH, Stripp A (1993) Codebreakers—the inside story of Bletchley Park. Oxford University Press, Reading
3. Fucks W (1965) Formulas to power. Forecasts about peoples, economy, potentials, 3rd edn. German Publishing House, Stuttgart
4. Mainzer K (2014) The computation of the world. From the theory of everything to big data. Beck, Munich
5. Planning Outline for the Construction for a Social Credit System (2014–2020) China Copyright and Media. 14 June 2014 (wordpress.com)
6. Yonxiang Lu (ed) (2010) Science & technology in China: a roadmap to 2050. Strategic general report of the Chinese academy of sciences, Beijing (Science Press). Springer, Heidelberg
7. Chunli B (ed) (2014) Vision 2020: emerging trends in science and technology and strategic option of China. BCAS 28(1):107–113
8. Confucius (2007) The Great Learning = Da Xue. Chinese Text Project. http://ctext.org/analects. Accessed 18 Jan 2016
9. Huang C (2009) Confucianism: continuity and development. transcript, Bielefeld
10. Hayes LD (2012) Political systems of East Asia. China, Korea and Japan. Routledge, New York
11. Hartmann WD, Maenning W, Wang Run (2017) China's new silk road. Cooperation instead of isolation—the role reversal in world trade. Frankfurter Allgemeine Buch, Frankfurt a. M.
12. Kant I (1900) Edition of the Prussian Academy of Sciences. Berlin, AA IV, 421: "Act only according to the maxim by which you can at the same time want it to become a general law."
13. Nida-Rümelin J (2011) Responsibility. Reclam, Stuttgart
14. Hart HL (1968) Punishment and responsibility. Essays in the philosophy of law. Oxford University Press, Oxford
15. Baumgartner HM, Eser A (eds) (1983) Guilt and responsibility: philosophical and legal contributions to the accountability of human actions. Mohr Siebeck, Tübingen, p 136

16. Zech H (2012) Information as object of protection. Mohr Siebeck, Tübingen
17. Mainzer K (2016) Information: algorithm, probability, complexity, quantum world, life, brain, society. Berlin University Press, Berlin, p 172
18. Jonas H (1979) The principle of responsibility. Attempt at ethics for technological civilization. Suhrkamp, Frankfurt

Printed in the United States
By Bookmasters